NEUROSCIENCE RESEARCH PROGRESS

WORKING MEMORY

DEVELOPMENTAL DIFFERENCES, COMPONENT PROCESSES AND IMPROVEMENT MECHANISMS

NEUROSCIENCE RESEARCH PROGRESS

Additional books in this series can be found on Nova's website under the Series tab.

Additional e-books in this series can be found on Nova's website under the e-book tab.

NEUROSCIENCE RESEARCH PROGRESS

WORKING MEMORY

DEVELOPMENTAL DIFFERENCES, COMPONENT PROCESSES AND IMPROVEMENT MECHANISMS

HELEN ST. CLAIR-THOMPSON
EDITOR

New York

Copyright © 2013 by Nova Science Publishers, Inc.

All rights reserved. No part of this book may be reproduced, stored in a retrieval system or transmitted in any form or by any means: electronic, electrostatic, magnetic, tape, mechanical photocopying, recording or otherwise without the written permission of the Publisher.

For permission to use material from this book please contact us:
Telephone 631-231-7269; Fax 631-231-8175
Web Site: http://www.novapublishers.com

NOTICE TO THE READER

The Publisher has taken reasonable care in the preparation of this book, but makes no expressed or implied warranty of any kind and assumes no responsibility for any errors or omissions. No liability is assumed for incidental or consequential damages in connection with or arising out of information contained in this book. The Publisher shall not be liable for any special, consequential, or exemplary damages resulting, in whole or in part, from the readers' use of, or reliance upon, this material. Any parts of this book based on government reports are so indicated and copyright is claimed for those parts to the extent applicable to compilations of such works.

Independent verification should be sought for any data, advice or recommendations contained in this book. In addition, no responsibility is assumed by the publisher for any injury and/or damage to persons or property arising from any methods, products, instructions, ideas or otherwise contained in this publication.

This publication is designed to provide accurate and authoritative information with regard to the subject matter covered herein. It is sold with the clear understanding that the Publisher is not engaged in rendering legal or any other professional services. If legal or any other expert assistance is required, the services of a competent person should be sought. FROM A DECLARATION OF PARTICIPANTS JOINTLY ADOPTED BY A COMMITTEE OF THE AMERICAN BAR ASSOCIATION AND A COMMITTEE OF PUBLISHERS.

Additional color graphics may be available in the e-book version of this book.

Library of Congress Cataloging-in-Publication Data

Library of Congress Control Number: 2013938561

ISBN: 978-1-62618-927-0

Published by Nova Science Publishers, Inc. † New York

CONTENTS

Preface		vii
Chapter 1	Working Memory and Children's Scholastic Attainment from 7 to 15 Years of Age: Developmental Differences and the Contribution of Speed of Processing *Helen St. Clair-Thompson*	1
Chapter 2	Difficulty of Children with Normal but Relatively Poor Working Memory at Japanese and Mathematics Classes *Masamichi Yuzawa and Miki Yuzawa*	17
Chapter 3	The Role of Working Memory and Inhibition in the Development of Motor Planning between 7 and 10 Years of Age *Valérie Pennequin*	31
Chapter 4	Contribution of Working Memory to Implicit Motor Learning in Children: A Preliminary Report *Alison Colbert and Jin Bo*	43
Chapter 5	The Development of Visuo-spatial Working Memory in Children *Colin Hamilton*	55
Chapter 6	Aging and Visual Feature Binding in Working Memory *Richard J. Allen, Louise A. Brown and Elaine Niven*	83
Chapter 7	Interactions between Cognitive Load Factors on Working Memory Performance in Laboratory and Field Studies *Edith Galy and Claudine Mélan*	97
Chapter 8	Affective Words Influence Processing in Visual and Auditory Working Memory *Fumiko Gotoh*	115
Chapter 9	Verbal and Visuo-spatial Processes in Spatial Orientation and Navigation *Andre Garcia and Carryl L. Baldwin*	131

Chapter 10	Working Memory Improvement by the Differential Outcomes Procedure *Ginesa López-Crespo and Angeles F. Estéve*	**145**
Chapter 11	Bilateral Transcranial Alternating Current Stimulation (tACS) of the Dorsolateral Prefrontal Cortex Enhances Verbal Working Memory and Promotes Episodic Memory After-Effects *Oded Meiron and Michal Lavidor*	**157**
Chapter 12	Cognitive Remediation of Working Memory in ADHD *Robert M. Roth and Shanna Treworgy*	**175**
Chapter 13	A Dual-Component Analysis of Working Memory Training *Bradley S. Gibson and Dawn M. Gondoli*	**201**
Index		**219**

PREFACE

Working memory is a limited capacity system for maintaining information during thinking and reasoning tasks. Some researchers use the term working memory interchangeably with short-term memory. However, commonly short-term memory refers to a system specialized for storing information whereas working memory refers to a system capable of processing in addition to storage. There are many theoretical conceptualizations of working memory which differ in their views of the nature, structure and function of working memory. However, despite these competing approaches it is widely accepted that working memory plays a central role in many aspects of cognition. Working memory has therefore been the focus of a huge amount of research, exploring its development and its cognitive underpinnings. There is also a rapidly growing literature examining cognitive changes that result from training on working memory tasks. This edited collection presents the latest research into working memory from around the world. It is concerned with three main themes; developmental differences, component processes, and improvement mechanisms.

The collection begins with chapters concerned with the role of working memory in children's development. Chapter 1 examines relationships between working memory and scholastic attainment throughout childhood. It is well-established that working memory plays an important role in children's learning and attainment, and that scores on working memory tasks serve as a useful predictor of literacy, mathematics and general school achievement. However, there have been suggestions that working memory is particularly important during the early school years, with later achievement being more dependent on higher level cognitive skills rather than working memory. The study reported in this chapter reveals that both the storage and executive-attentional components of working memory remain closely related to children's attainment from 7 to 15 years of age. However, it does reveal developmental differences in the contributions made by memory storage and speed of processing. Speed of processing appeared to be uniquely linked with attainment from 11-13 years of age only, which may be a result of the assessment used for children's scholastic attainment.

Chapter 2 then explores the difficulties that children with normal but relatively poor working memory have in the school classroom. This particular participant group was chosen to demonstrate that the difficulties children have in school are caused by a mismatch between their working memory skills and teachers' pedagogical approaches, rather than children's working memory *per se*. In the studies reported in this chapter it was found that children with relatively poor working memory rarely raised their hands when teachers asked questions, and did not pay attention when peers were speaking, therefore finding it difficult to contribute to class discussions. Further research revealed that these difficulties are relatively enduring. The results are discussed in terms of implications for educational practice. For example, strategies to encourage children to raise their hands in class are described. In particular the chapter emphasises that children with a poor working memory require appropriate support in the school classroom.

Moving on from examining associations between working memory and children's learning and classroom performance, chapter 3 explores the role of working memory and also the inhibition of prepotent responses in the development of the executive process of planning in children. Planning is an important cognitive function that guides responses and enables successful goal-directed behaviour. The study reported in this chapter used an inventive methodology allowing working memory load and inhibitory requirements to be manipulated within a planning task. The results revealed an interaction between working memory and response inhibition such that taxing working memory resources increased the difficulty of inhibiting prepotent responses in planning activities. This supported previous findings of a close link between working memory and inhibition. The findings also revealed roles for working memory and inhibition in planning. However, there were developmental changes in the success of planning, but no developmental difference in the interaction between working memory load and inhibition. The authors consider the importance of examining several measures, including planning time and execution time in addition to planning performance, and the importance of children's planning strategies is also discussed. It is suggested that further research is needed to examine associations between working memory, inhibition and planning.

Chapter 4 then examines the role of working memory in children's implicit motor learning. It is well established that working memory processes are important in explicit learning, as described in the earlier chapters. However, this chapter explores the possibility that working memory is also important in implicit learning, during which participants are not consciously aware that learning has occurred. In the study described in this chapter children were tested on verbal and visuo-spatial change detection working memory tasks and analyses examined the correlations between working memory and performance on a serial reaction time task assessing implicit learning. There was no correlation between working memory and implicit motor sequence learning. The findings are interpreted within a developmental invariant hypothesis, suggesting that implicit learning does not develop steadily over childhood, and instead reaches adult competence early in life. The chapter concludes that there is only a minor role for working memory in implicit sequence learning in typically developing children.

The final chapter concerned with the development of working memory in children is Chapter 5. This is concerned specifically with the development of visuo-spatial working memory. It provides a detailed review of the literature concerned with children's employment of visual, spatial mnemonic, and attentional processes during visual working memory tasks. It is suggested that a diverse range of cognitive resources may be employed during these tasks, and that their employment is partly dependent upon children's age, with a developmental increase in the use of strategies and semantics to support working memory task performance. However, the author also suggests that the employment of strategies is dependent upon the demands of working memory tasks, and therefore recommends using a task analysis approach when examining the development of working memory. The research within this chapter is embedded within well-established theoretical approaches to working memory. This therefore leads into the second main theme of the book, the component processes of working memory.

Chapter 6 provides a detailed review of research on the process of binding, referring to how features of a stimulus are bound together in working memory to form an integrated representation. Binding has important implications for theoretical models of working memory, but also for understanding the development of brain disorders. The research reported in the chapter demonstrates that in healthy adults the process of binding is not attentionally demanding, but that bound representations are particularly fragile and open to interference. Binding is then examined in terms of aging effects and Alzheimer's disease. In healthy aging there appears to be inconsistent findings regarding a binding deficit, which may be in part a consequence of differences in methodology and studies assessing different forms of information binding. In Alzheimer's disease binding deficits appear to be more robust, with many studies demonstrating working memory binding deficits. It is suggested that further research is needed to develop a clearer understanding of what conditions give rise to age-related binding deficits, and that this would further our understanding of cognitive aging as well as Alzheimer's disease.

Chapter 7 then examines working memory from a cognitive load perspective. Cognitive load theory distinguishes between intrinsic load (induced my material to be processed), extraneous load (induced by external factors such as time pressure), and germane load (referring to the application of task appropriate strategies). The studies presented in the chapter examine the interactive effects of intrinsic, extraneous and germane load on working memory in both laboratory and field studies. Intrinsic load (task difficulty) and extraneous load (time pressure) are found to have additive effects on working memory performance, such that performance is poorest in conditions of high difficulty with added time pressure. However, interestingly, the results also suggest effects of alertness, with effects of intrinsic and extraneous load being emphasised when cognitive resources are reduced due to low levels of alertness. The findings have implications for models of cognitive load and the authors conclude that there is a need for further studies to address cognitive load theory and working memory requirements in the field of ergonomics.

Chapter 8 explores the topic of how emotional or affective stimuli influence processing in working memory. The chapter provides a review of a series of studies concerned with affective words in both visual and auditory domains. In both domains affective stimuli are shown to induce dominance, interference, and facilitation effects when compared to neutral stimuli. Importantly, the results are attributed to a central attention mechanism of working memory. It is suggested that affective stimuli capture this attention mechanism, either facilitating or interfering with on-going processing, independently of biases related to anxiety, and also independently of arousal.

Chapter 9 then provides a review of research concerned with working memory and spatial orientation and navigation. It seems that individual differences in spatial orientation and navigation are indeed associated with working memory or working memory strategies. For example, individuals with a poor sense of direction rely on verbal working memory during navigation, whereas individuals with a good sense of direction employ visuo-spatial working memory. Individuals with a poor sense of direction are also required to use more working memory resources during orientation and navigation tasks. The findings presented in the chapter have important implications, for example for the design of navigational systems.

The third main theme of this collection, improvement mechanisms for working memory, then begins with chapter 10. This chapter provides a review of studies concerned with the differential outcomes effect, the improvement of conditional discrimination learning when alternative choice responses are followed by different reinforcers. The studies discussed relate to the theoretical basis of the effect, its biological underpinnings, and its potential as a method to improve memory. Of particular interest are a series of studies demonstrating that the differential outcomes procedure can enhance working memory, for example in Alzheimer's disease, in old adults, and in children born prematurely. The procedure therefore has important implications for improving working memory.

Chapter 11 discusses the possibility of improving working memory performance using transcranial alternating current stimulation (tACS). The study reported in this chapter revealed that working memory was significantly improved when tACS was directed to the dorsolateral prefrontal cortex (DLPFC). Explicit episodic retrieval, assessed through delayed free recall, was also enhanced. It was concluded that bilateral DLPFC alternating stimulation can be used to increase functional connectivity in order to improve working memory and episodic memory formation. These findings have important implications, particularly for clinical populations that suffer from loosened theta-band functional connectivity, for example schizophrenia patients. DLPFC tACS could be used as a therapeutic application to enhance working memory in these patients.

In Chapter 12 the topic of improving working memory then turns to Attention deficit hyperactivity disorder (ADHD). This chapter provides a comprehensive review of research concerned with the effects of medication and behavioural approaches on improving working memory, before focussing on cognitive remediation employing well-known training regimes such as Cogmed training. Findings are discussed with regards to the effects of training on near-transfer, far-transfer, and the core symptoms of ADHD. Methodological issues are also considered, including study design, nature of training programs, and expectancy effects. Although the results of cognitive training studies in participants with ADHD are varied, particularly regarding far-transfer, it is concluded that cognitive remediation is a promising approach for improving working memory in ADHD.

The topic of cognitive remediation of working memory via training is also addressed in Chapter 13. Some previous research has been optimistic about the possibility that training can enhance working memory and related skills, but several recent reviews have been much more pessimistic, concluding that there is little evidence of benefits, and particularly of transfer. However, the authors of this chapter suggest that previous findings of weak or ineffective training of working memory and associated abilities may be a result of training programs not being optimally designed in relation to theoretical models of working memory. Evidence is presented that suggests that previous training studies have only targeted primary, or short-term storage components of working memory, and not the ability to retrieve information from secondary memory. A newly created regime based on decreasing the recall accuracy threshold is then described, which also targets secondary memory or executive-attentional resources. The success of this new training technique for improving working memory and associated abilities still needs to be fully examined. However, it is likely to be influential in many future working memory training studies.

Together, the chapters in this collection add to a growing literature concerned with working memory. Regarding development, the chapters reiterate the importance of working memory in children's learning and school performance and the need for educators to consider the working memory demands of classroom activities. Working memory is important not only for the development of scholastic skills, but also for the development of functions like planning, which plays a major role in determining goal-directed behaviour, and also spatial navigation and orientation, common and important tasks of everyday life.

The research presented can also inform readers about the component processes of working memory. Much research into working memory is focused upon executive or attentional resources. One process that has important implications for understanding cognitive aging is that of visual feature binding. Research presented in this collection appears to suggest, that at least in some participants, binding occurs relatively independently of executive processes and thus may occur early on in visual selection. Elsewhere in the collection it is shown that executive resources appear to be allocated to affective stimuli entering working memory, and that this allocation appears to be relatively automatic, either interfering with or facilitating further processing.

The research described in several chapters of this collection also furthers our understanding of potential improvement mechanisms for working memory. It appears that in some clinical groups working memory can be improved by employing the differential outcomes procedure, or by using tACS of the DLPFC. Cognitive remediation of working memory through working memory training may also be of benefit to these patient groups, as well as to individuals with ADHD, a poor working memory, or even typically developing children.

There are also a number of important themes that emerge throughout the chapters which are perhaps not captured in the chapter titles. For example, although there are many theoretical approaches to working memory, many of which make distinctions between storage and executive-attentional control (or primary and secondary memory) the research presented here demonstrates that is it important for researchers to consider the demands of working memory tasks and thus other mechanisms that may potentially play a role in working memory performance. Chapter 5 describes how a range of processes, including strategies and semantics contribute to working memory task performance, and suggests that only some of the major theories of working memory can account for these findings. Similarly, Chapter 13 discusses the importance of considering the demands of working memory training activities, particularly whether they employ primary or secondary memory, before a conclusion about the malleability of working memory can be reached. Chapter 12 suggests that when considering working memory performance we should also be aware of other factors which can affect the availability of working memory resources, such as alertness or time pressure, or the knowledge of appropriate task strategies. The use of strategies is also a theme embedded in chapter 9, which describes how individual differences in spatial orientation and navigation may relate to working memory resources, for example the employment of verbal or visuo-spatial strategies. Chapter 3 also states that it is important to consider strategies in relation to children's planning performance.

The research presented in this collection therefore furthers our knowledge about working memory, in terms of its development, component processes, and improvement mechanisms, but also demonstrates how further research is needed in order to develop a full understanding of the cognitive resources that underlie working memory. A more detailed understanding of these resources can enhance research into the development of working memory, and also potential methods of improving working memory and associated skills. It is hoped that the research presented here will spark an interest in developing this understanding and that it will form the basis for much future research into working memory.

In: Working Memory
Editor: Helen St. Clair-Thompson

ISBN: 978-1-62618-927-0
© 2013 Nova Science Publishers, Inc.

Chapter 1

WORKING MEMORY AND CHILDREN'S SCHOLASTIC ATTAINMENT FROM 7 TO 15 YEARS OF AGE: DEVELOPMENTAL DIFFERENCES AND THE CONTRIBUTION OF SPEED OF PROCESSING

Helen St. Clair-Thompson
University of Hull, Hull,UK

ABSTRACT

It is well established that working memory is predictive of learning and attainment during childhood. However, there may be developmental differences in the relationships between working memory and attainment. Some research has suggested that the correlations between scores on working memory tasks and attainment decrease throughout childhood, as achievement becomes more dependent upon higher-level conceptual and analytical abilities. In contrast, other studies have revealed robust links between working memory and attainment throughout childhood and adulthood. The current study therefore examined links between the phonological loop, visuo-spatial sketchpad, and central executive components of working memory and children's scholastic attainment from 7 to 15 years of age. Scores on measures of each component of working memory remained significantly associated with attainment in each year group. The study also explored the cognitive underpinnings of relationships between working memory and attainment. Thus processing speeds were recorded in addition to memory scores for the measure of the central executive. The results revealed interesting developmental differences in the relationships between speed of processing and attainment. At younger ages processing speed was significantly related to attainment, but did not predict any unique variance over and above memory span. With increasing age the unique contribution of speed of processing increased, up to aged 13, after which point speed made no contributions to attainment. The results are discussed in terms of both theoretical and practical implications. For example, if the purpose of using working memory tasks is to predict attainment they suggest benefits of including measures of speed of processing in assessments of working memory for some age groups.

INTRODUCTION

Working memory is a temporary storage system responsible for storing and manipulating information during complex and demanding activities (Baddeley, 1986). It underlies our capacity for complex thought (Baddeley, 2007). There are several theoretical models which differ in their views of the nature, structure, and function of the working memory system (e.g. for a review see Conway, Jarrold, Kane, & Towse, 2007; Miyake & Shah, 1999). However, one widely accepted model (Baddeley, 2000; Baddeley & Hitch, 1974) proposes that working memory consists of four components. There are two domain-specific short-term memory components; the phonological loop that is responsible for the maintenance of verbal information, and the visuo-spatial sketchpad that is responsible for dealing with visual and spatial information. These are governed by a central executive system, a domain-general limited capacity system often likened to a mechanism of attentional control (e.g. Engle & Kane, 2004; Engle, Kane & Tuholski, 1999; Kane & Engle, 2003; Unsworth & Engle, 2007). Baddeley (2000) identified the episodic buffer as a further subcomponent of working memory, forming an interface between the subcomponents of working memory and long-term memory.

The subcomponents of working memory can be assessed using a range of measures. The phonological loop is commonly assessed using tasks which involve the immediate recall of verbal information, for example word recall or digit recall, and the visuo-spatial sketchpad is typically assessed using tasks which involve the immediate recall of visual or spatial information, such as pattern recall or block recall (e.g. Pickering & Gathercole, 2001). The central executive can be assessed using a range of measures (e.g. see Oberauer, 2005), but amongst the most commonly used measures are working memory span measures requiring both the processing and storage of information. For example, in counting span (Case, Kurland, & Goldberg, 1982) participants count the number of items in a series of arrays and then recall the successive tallies of each array. Within the multiple component model of working memory (Baddeley, 2000; Baddeley & Hitch, 1974) it is thought that processing employs the central executive and maintenance involves the domain specific storage systems (e.g. Baddeley & Logie, 1999; Duff & Logie, 2001; LaPointe & Engle, 1990). Tasks assessing the storage components and the central executive of working memory therefore overlap in terms of their requirements. They both require the maintenance of information, and may both invoke controlled processes such as rehearsal (e.g. Engle, Tuholski, Laughlin, & Conway, 1999; Unsworth & Engle, 2007). However, in tasks assessing the central executive controlled processing is required in the context of high demands upon attention or maintaining information in the face of interference (e.g. Engle et al., 1999).

Over recent years working memory has increasingly been studied within an educational context. In the typical school classroom many demands are placed upon the working memory system. There are many activities that impose simultaneous processing and storage demands, including listening whilst trying to take notes, following instructions, decoding unfamiliar words, and mental arithmetic. In each of these activities a learner must process information and integrate it with previously stored knowledge or recently encountered information. There are therefore many opportunities for children to fail on classroom tasks as a result of a poor working memory. Children with poor working memory often lose track of their place in complex tasks, require frequent repetition of instructions and important information, skip procedural steps, and often abandon a task before completion (e.g. Alloway & Gathercole, 2006).

Working memory is therefore highly predictive of a number of scholastic skills during childhood, including literacy, mathematics and comprehension (e.g. Alloway & Passolunghi, 2011; Bull & Scerif, 2001; Cain, Oakhill & Bryant, 2004; De Jong, 1998; De Stefano & LeFevre, 2004; Engel de Abreu, Gathercole & Martin, 2011; Seigneuric, Ehrlich, Oakhill & Yuill, 2000; Swanson & Kim, 2007). Working memory is consequently closely related to children's performance on national curriculum assessments of English, mathematics, and science in England (Gathercole, Brown, & Pickering, 2003; Gathercole & Pickering, 2000; Gathercole, Pickering, Knight, & Stegmann, 2004; Jarvis & Gathercole, 2003; St Clair-Thompson & Gathercole, 2006) and is also related to teacher ratings of children's academic performance in these areas (e.g. St Clair-Thompson & Sykes, 2010).

Each of the subcomponents of the multiple-component model of working memory (Baddeley, 2000; Baddeley & Hitch, 1974) shares close links with learning and attainment. The phonological loop has been associated with learning vocabulary in both native and foreign languages (e.g. Engel de Abreu, Gathercole, & Martin, 2011; Majerus, Poncelet, Greffe, & van der Linden, 2006; Masoura & Gathercole, 2005). It has thus been argued that an important role of the phonological loop is to temporarily store information whilst stable representations of words are formed in long-term memory (Gathercole, 2006). The phonological loop also shares links with mathematical problem solving (e.g. Alloway & Passolunghi, 2011; Seyler, Kirk, & Ashcraft, 2003), which have been attributed to its role in maintaining operands and interim results during calculations (e.g. Alloway & Passolunghi, 2011; Seitz & Schumann-Hengsteler, 2000). The visuo-spatial sketchpad has been related to arithmetic and mathematics (e.g. Alloway & Passolunghi, 2011; Holmes, Adams & Hamilton, 2008), functioning as a mental blackboard to support number representation, or to encode problems that are presented visually (e.g. D-Amico & Guarnera, 2005; Trbovich & LeFevre, 2003). Finally, the central executive system has been associated with learning and attainment across a number of domains, including syntax and reading development (e.g. Engel de Abreu et al., 2011), arithmetic abilities and mathematical performance (e.g. Alloway & Passolunghi, 2011; Swanson & Kim, 2007), and overall academic attainment (Alloway & Alloway, 2010; St Clair-Thompson & Gathercole, 2006).

Working memory deficits have also been implicated in many learning difficulties (e.g. Alloway, Gathercole, Willis, & Adams, 2005). For example, children with specific language impairment typically perform poorly on measures of the phonological loop and central executive component of working memory (e.g. Archibald & Gathercole, 2006; Alt, 2011). Similarly, children with dyslexia commonly perform poorly on measures of the phonological loop and central executive but not measures of the visuo-spatial sketchpad (e.g. Jeffries & Everatt, 2004; Kibby, Marks, Morgan, & Long, 2004; but see Menghini, Finzi, Carlesimo, & Vicari, 2011). Working memory deficits have also been associated with neurodevelopmental disorders such as ADHD (e.g. Alloway, 2011; Sowerby, Deal & Tripp, 2011).

It is, however, important to note that there may be developmental differences in the relationships between working memory and learning and attainment. Jarrold, Baddeley, Hewes, Leeke, and Philips (2004) investigated the relationships between the phonological loop and vocabulary acquisition in childhood. They demonstrated that verbal short-term memory was causally related to individual differences in vocabulary acquisition, but only early on in development. The link between the phonological loop and vocabulary acquisition in later years may have disappeared due to the dominance of other factors such as general intelligence and degree of language exposure. Imbo and Vandierendonck (2007) examined the contribution of the central executive to simple arithmetic. Their results suggested that as children grow older, and are presumably more experienced with solving simple addition problems, they become more efficient at retrieving information and using counting strategies, resulting in a reduced need for executive resources (see also Raghubar, Barnes & Hecht, 2010). Is also appears that the visuo-spatial sketchpad may be implicated to a greater degree in the maths performance of younger children who are in the process of acquiring basic arithmetic skills (e.g. Rourke, 1993), but that with experience verbal working memory comes to support arithmetic to a greater extent. Altemeier, Jones, Abott, and Berninger (2006) further posited that executive functions may be more important earlier in schooling, when academic skills are less automatic and require more effortful planning and controlled processing.

Research has also suggested developmental differences in the relationship between working memory and overall school attainment. Gathercole et al. (2004) examined relationships between the phonological loop and central executive components of working memory and children's national curriculum levels at 7 and 14 years of age. Strong links were found between the central executive and English attainment at aged 7 but not at aged 14. Gathercole et al. further noted that at aged 7 the tasks used for English directly tapped literacy abilities, whereas at aged 14 they tapped higher-level conceptual and analytic abilities. They therefore concluded that working memory is particularly important for the acquisition of literacy skills, but is not so important for higher level language and reasoning. Other researchers, however, have suggested that the central executive remains important for language and literacy throughout the childhood years. For example, Jarvis and Gathercole (2003) found significant relationships between working memory and attainment in English, mathematics and science at both 11 and 14 years of age (see also St Clair-Thompson & Gathercole, 2006).

The first aim of the current study was therefore to examine the relationships between working memory and attainment throughout childhood, with a view to identify any developmental differences. The study involved a large number of participants who all completed the same assessments of the phonological loop, visuo-spatial sketchpad and central executive components of working memory. Schools supplied the latest national curriculum attainment levels for each child.

The second aim of the current study was to explore the cognitive underpinnings of the relationships between working memory and scholastic attainment. Although each of the subcomponents of the multiple-component model of working memory (Baddeley, 2000; Baddeley & Hitch, 1974) shares close links with children's learning and attainment, many studies have suggested that performance on complex span tasks tapping the central executive are more closely associated with measures of attainment than scores on short-term memory tasks (e.g. Daneman & Carpenter, 1980; Engle et al., 1999). Complex working memory span tasks involving the central executive are dual task measures. Thus, there are at least two sources of data, one from processing and one from storage. There are several theoretical models of complex span task performance which assume close relationships between processing and storage (e.g. Barrouillet, Bernadin, & Camos, 2004; Case et al., 1982; Towse, Hitch, & Hutton, 2000). However, some studies have revealed only a minor association between the time taken to complete processing activities and the accuracy of recall on working memory tasks (e.g. Unsworth, Heitz, Schrock, & Engle, 2005). Processing and storage have also been found to contribute unique variance to complex span task performance (e.g. Bayliss, Jarrold, Gunn, & Baddeley, 2003; see also Bayliss, Jarrold, Baddeley, Gunn & Leigh, 2005).

Researchers have therefore explored the interrelationships between processing and storage in working memory tasks and higher order cognition. Processing durations have been found to predict unique variance in higher order cognition over and above working memory span scores. For example, Waters and Caplan (1996) found that the addition of processing times to reading span scores increased the correlation between reading span and comprehension. Similarly, Bayliss et al. (2003) demonstrated that processing times were a significant predictor of reading and mathematics in children when statistically controlling for storage capacity (see also Cowan, Towse, Hamilton, & Saults et al., 2003; Hitch, Towse, & Hutton, 2001). The second aim of the current study was therefore to explore the relative contributions of processing and storage in working memory to children's educational attainment.

Therefore in addition to children completing measures of the phonological loop, visuo-spatial sketchpad and central executive components of working memory, for the central executive measure both memory scores and processing times were recorded in the current study. The relationships between memory scores, speeds of processing, and educational attainment were explored in children aged 7 to 15 years of age.

METHOD

Participants

The participants were 604 children aged 7 to 15 years. The children were separated into groups based on chronological age. The number of children in each group is detailed in Table 1. The children were recruited from several schools in the North of England. Parental consent was obtained for each child prior to testing. No information regarding parental income or educational level, family size, or the ethnicity of children was collected.

Table 1. Participant information

	Total N	Males	Females	Mean age
Age 7	50	24	26	7 years 3 months
Age 8	69	37	32	8 years 6 months
Age 9	71	35	36	9 years 5 months
Age 10	81	38	43	10 years 6 months
Age 11	66	33	33	11 years 4 months
Age 12	106	56	50	12 years 3 months
Age 13	57	29	28	13 years 4 months
Age 14	55	26	29	14 years 6 months
Age 15	49	24	25	15 years 3 months

Materials and Procedure

Participants completed a new computerised assessment of working memory; Lucid Recall (St Clair-Thompson, 2013). This assessment includes three measures of memory, designed to tap the phonological loop, visuo-spatial sketchpad, and central executive components of working memory.

The phonological loop task was word recall. Participants were asked to recall, in the same order, sequences of monosyllabic words presented aloud through the computer headphones. The words were presented at the rate of one per second. Following completion of the sequence participants had to use the computer mouse to select the targets from amongst a number of distracters. After two practice trials, testing began with a maximum of six trials with one word. The number of words was then increased by one if a child successfully recalled the words in four trials at a list length. When three or more lists of a particular length were recalled incorrectly testing ended. The score was calculated as the number of trials on which the words were recalled correctly. Test-retest reliability for this task has been calculated as .71 for children aged 7-9 years and .68 for children aged 10-12 years (St Clair-Thompson, submitted).

The visuo-spatial sketchpad task was pattern recall. Participants were presented with a series of matrix patterns, for 2 seconds each. Following presentation of each pattern participants provided their response by using the computer mouse to click in a blank matrix to recreate the pattern. Following two practice trials, testing began with a maximum of six trials with two filled squares in the matrix. The number of filled squares was then increased by one if a child successfully recalled the pattern in four trials of one matrix size. When three or more patterns within a block were recalled incorrectly testing ended. The score was calculated as the number of trials on which the pattern was recalled correctly. Test-retest reliability for this task has been calculated as .69 for children aged 7-9 years and .77 for children aged 10-12.

The central executive task was counting recall, in which participants counted the number of items in a series of arrays and then recalled the successive tallies of each array. The target items were red circles, which were presented amongst distracters which were red squares and blue circles. Participants responded to each trial by using the computer mouse to select the correct response from the numbers 1 to 9, and then used the same procedure to recall the tallies at the end of each trial. The structure of testing, discontinuation criteria and scoring method were the same as for the digit recall and pattern recall tasks described above, with the exception that the computer also recorded a mean reaction time for the counting operations for each participant. Test-retest reliability for this task has been calculated as .59 for children aged 7-9 years and .76 for children aged 10-12.

All children completed the working memory tasks in the computer classroom in school. The memory assessment software was loaded on to the school computers by the computer technician. Children then entered the classroom on a class basis (the largest class was comprised of 28 pupils). Each child sat at a computer and was given instructions of how to enter their name and date of birth. They were then instructed to wear the headphones which had been provided, and to listen to the instructions given by the computer. The tasks were then automated, so no further experimenter involvement was required.

The schools supplied National Curriculum attainment levels in reading, writing, and mathematics for each pupil, although for children aged 14 and 15 years only levels for

English and mathematics were available. Levels comprised teacher's assessments of children's progress on the National Curriculum, measured by tasks and tests that are administered informally. Previous research has revealed that such teacher ratings show correlations with working memory, similar to when using standardised test scores (e.g St Clair-Thompson & Sykes, 2010).

RESULTS

The mean scores on the working memory tasks for each age group are shown in Table 2. For each working memory task there was a developmental increase in performance. The only exception was word recall, for which there was no improvement between ages 14 and 15. There was also a developmental increase in counting speeds, with older children responding more quickly than younger children.

Table 2. Mean scores on the working memory tasks (*Standard Deviations* are shown in parentheses)

	Word recall	Pattern recall	Counting recall	Counting speed
Age group				
7	8.98 (4.45)	20.41 (8.55)	7.22 (4.35)	5011.87 (1549.55)
8	10.53 (4.50)	23.72 (10.31)	10.05 (5.58)	4369.30 (1247.48)
9	10.93 (5.34)	25.04 (10.52)	10.28 (6.00)	4113.16 (1435.39)
10	11.90 (5.30)	28.89 (9.74)	11.05 (4.76)	3683.71 (1235.39)
11	14.89 (5.04)	32.12 (9.16)	13.32 (6.24)	3193.71 (1128.44)
12	16.20 (4.20)	32.54 (6.93)	13.62 (6.20)	2953.67 (1212.27)
13	17.64 (3.47)	33.29 (6.68)	14.25 (5.70)	2947.36 (1034.82)
14	18.74 (4.38)	38.64 (6.95)	18.93 (7.21)	2890.53 (810.26)
15	17.51 (6.18)	38.93 (9.26)	20.00 (7.35)	2820.59 (620.08)

Table 3 shows the correlations between working memory and attainment in reading, writing, and mathematics (or English and mathematics) for each age group. With a small number of exceptions (pattern recall and reading at aged 12, and word recall and counting recall with mathematics at aged 14) scores on each measure of working memory were significantly related to attainment from 7 to 15 years of age. Table 3 also shows the relationships between speed of processing during the counting recall task and attainment in each curriculum area. Speed was not significantly related to attainment at aged 7. However, speed was significantly related to attainment from 8 to 15 years of age. The relationship was negative, suggesting that faster speed of processing is associated with higher levels of attainment.

Counting recall scores and counting speeds were also significantly negatively correlated between aged 8 and aged 15, with correlations of -.33, -.36, -.37, -.29, -.24, -.30, -.27, and -.29 ($p < .05$ in each case). At aged 7 counting recall and counting speed were not significantly related, with a correlation of -.21 ($p > .05$).

Table 3. Correlations between working memory and attainment in each age group

		Word recall	Pattern recall	Counting recall	Counting speed
Age 7					
	Reading	.46**	.36*	.52**	-.12
	Writing	.31*	.29*	.40**	-.14
	Mathematics	.41**	.34*	.52**	-.21
Age 8					
	Reading	.45**	.45**	.58**	-.32*
	Writing	.35**	.41**	.50**	-.28*
	Mathematics	.46**	.43**	.56**	-.28*
Age 9					
	Reading	.45**	.35**	.48**	-.35**
	Writing	.43**	.44**	.60**	-.39**
	Mathematics	.31**	.37**	.48**	-.36**
Age 10					
	Reading	.50**	.46**	.45**	-.31**
	Writing	.58**	.52**	.39**	-.40**
	Mathematics	.45**	.47**	.45**	-.23*
Age 11					
	Reading	.47**	.46**	.31*	-.65**
	Writing	.50**	.55**	.32*	-.58**
	Mathematics	.40**	.41**	.32**	-.42**
Age 12					
	Reading	.34**	.19	.29**	-.31**
	Writing	.47**	.22**	.40**	-.37**
	Mathematics	.32**	.20*	.24*	-.29**
Age 13					
	Reading	.38**	.29**	.30*	-.59**
	Writing	.35*	.27*	.41**	-.54**
	Mathematics	.42**	.45**	.45**	-.46**
Age 14					
	English	.31**	.39**	.46**	-.34*
	Mathematics	.23	.50**	.25	-.18
Age 15					
	English	.64**	.47**	.48**	-.29*
	Mathematics	.62**	.53**	.58**	-.29*

Note: ** p < .01, * p < .05

In order to identify whether counting speed and counting recall predicted unique variance in scholastic attainment a series of partial correlation analyses were then conducted. Correlations between counting recall scores and attainment were calculated whilst statistically controlling for speed of processing. Correlations between speed of processing and attainment were then calculated whilst partialling out counting recall scores. The results of the partial correlations are presented in Table 4. At aged 7, 8, and 9 memory scores, but not speed of processing, predicted unique variance in attainment. At aged 10, 11, 12, and 13 both memory scores and speed of processing predicted unique variance (for some scholastic domains). At aged 14 and 15, memory scores predicted unique variance but speeds of processing did not.

Table 4. Partial correlations between counting recall, counting speed, and attainment in each age group

		Reading	Writing	Maths
Age 7				
	Counting recall	.51**	.39**	.50**
	Counting speed	.01	-.06	-.12
Age 8				
	Counting recall	.53**	.45**	.51**
	Counting speed	-.16	-.14	-.10
Age 9				
	Counting recall	.40**	.53**	.40**
	Counting speed	-.21	-.22	-.21
Age 10				
	Counting recall	.39**	.29**	.41**
	Counting speed	-.19	-.31**	-.10
Age 11				
	Counting recall	.31**	.34**	.35**
	Counting speed	-.38**	-.35**	-.15
Age 12				
	Counting recall	.22*	.34**	.18
	Counting speed	-.26*	-.30**	-.24*
Age 13				
	Counting recall	.16	.31*	.37**
	Counting speed	-.55**	-.47**	-.35**

		English	Maths
Age 14			
	Counting recall	.40**	.21
	Counting speed	-.25	-.12
Age 15			
	Counting recall	.44**	.54**
	Counting speed	-.17	-.16

*Note: * p < .05 ** p < .01*

DISCUSSION

The current study had two main aims. The first was to examine the relationships between working memory and attainment throughout childhood, with a view to identify any developmental differences. The second aim was to explore the cognitive underpinnings of the relationships between working memory and attainment, by examining the relative contributions of memory storage and speed of processing.

The results of the study showed close relationships between working memory and attainment, across the domains of reading, writing and mathematics (or English and mathematics), throughout the childhood years. The findings are therefore consistent with previous studies that have demonstrated close relationships between working memory and children's performance on national curriculum assessments of English, mathematics, and science in England (Gathercole et al., 2003; Gathercole & Pickering, 2000; Gathercole et al., 2004; Jarvis & Gathercole, 2003; St Clair-Thompson & Gathercole, 2006; St Clair-Thompson & Sykes, 2010), and further suggest, contrary to Gathercole et al. (2004), that there is no developmental reduction in the relationships between working memory and achievement (see also Jarvis & Gathercole, 2003; St Clair-Thompson & Gathercole, 2006).

It is, however, important to note that national curriculum levels, used as measures of attainment in the current study, are general measures of attainment likely to assess a number of skills rather than specific scholastic competencies. Jarrold et al. (2004) demonstrated that the phonological loop became less important for vocabulary throughout development, and similarly Imbo and Vandierendonck (2007) demonstrated that the central executive became less necessary for arithmetic. However, the broad assessments used for the national curriculum are likely to rely upon many more abilities than vocabulary and arithmetic. For example, mathematics assessments at a young age require skills including applying mathematics, shapes, space and measures, and handling data. At an older age mathematics assessments may also require problem solving, ratios, fractions and algebra. These skills may continue to employ working memory resources throughout childhood and even into adulthood (e.g. Destefano & LeFevre, 2004; Raghubar et al., 2010). Therefore although working memory remains closely associated with scholastic attainment throughout childhood, the specific role played by working memory resources may well change throughout development.

It is interesting to note that close relationships were observed between each of the phonological loop, visuo-spatial sketchpad, and central executive components of working memory and children's attainment in each curriculum domains. Many studies have suggested that performance on complex span tasks tapping the central executive are more closely associated with measures of attainment than scores on short-term memory tasks tapping the phonological loop or visuo-spatial sketchpad (e.g. Daneman & Carpenter, 1980; Engle et al., 1999). However, again this may be due to national curriculum levels being fairly broad assessments relying on a wide range of skills and abilities. Although the central executive may play a more important role than the storage components of working memory in certain skills such as syntax and reading (e.g. Engel de Abreu et al., 2011), or mathematical performance (e.g. Alloway & Passolunghi, 2011; Swanson & Kim, 2007), the phonological loop is closely related to vocabulary (e.g. Engel de Abreu et al., 2011) and mathematical problem solving (e.g. Seyler et al., 2003), and the visuo-spatial sketchpad has been implicated in aspects of mathematics (e.g. Alloway & Passolunghi, 2011; Holmes et al., 2008). Therefore differing aspects of national curriculum assessments are likely to depend upon different subcomponents of the working memory system.

Regarding the second aim of the current study, to examine the relative contributions of memory storage and speeds of processing to scholastic attainment, the relationship between speed and attainment showed an interesting developmental pattern. Speed was significantly related to counting recall scores from aged 8 to aged 15, consistent with many theoretical models of working memory which emphasise the importance of processing speed (e.g. Barrouillet et al., 2004; Case et al., 1982; Towse et al., 2000). Speed and counting recall were not significantly correlated at aged 7, although this may in part reflect the smaller number of participants in this age group. Speed was also not significantly related to attainment at aged 7, but then was related to attainment from aged 8 until aged 15. Regarding the unique contribution of speed (whilst controlling for memory storage), speed did not predict attainment at aged 7, 8 or 9. It then did predict attainment at aged 11-13 (in addition to writing at aged 10). Speed no longer predicted unique variance at age 14 or 15. These results are consistent with previous studies which have revealed that processing times are significant predictors of attainment in children whilst statistically controlling for storage capacity (e.g. Bayliss et al., 2003, Cowan et al., 2003; Hitch et al., 2001; Waters & Caplan, 1996), but further suggest that this is only the case in certain age groups.

It is important to note that these developmental differences in the relationships between speed of processing and attainment were not due to developmental changes in the relationship between speed and memory scores. Rather, they may well be a result of the assessments used to measure scholastic attainment in each group. Speed of processing within working memory appeared to play an important unique role in attainment between ages 11 and 13. Children at this age are at Key Stage 3 of the national curriculum in England. The national curriculum dictates what children are taught and also how they are assessed. The findings of the current study would appear to suggest that assessments at this key stage place particular demands upon speed of processing in working memory, in addition to demands upon storage which occur throughout the childhood years. At this stage assessments are designed not only to directly tap literacy or numeracy abilities, but to assess higher-level conceptual and analytic abilities. It is possible that speed of processing in working memory is particularly important for these skills. However, further research is needed to establish why speed of processing is then not uniquely related to attainment at age 14 or 15.

It is, however, worthy of note that the speed of processing measure used in the current study refers only to the speed of counting the number of items in the counting recall task. Some researchers have used much more comprehensive measures of speed of processing, and some have calculated speeds of processing whilst participants were not required to simultaneously remember information (e.g. Bayliss et al., 2003). Therefore further research is needed to establish the robustness of the findings of the current study and also to explore relationships between memory storage, speeds of processing and attainment using different working memory and attainment measures.

There is, however, no question about the robustness of the relationships between working memory and attainment. A huge literature, using different working memory tasks and also different attainment measures, has revealed close relationships between working memory and scholastic skills (e.g. Alloway & Passolunghi, 2011; Engel de Abreu et al., 2011; Gathercole et al., 2004; Seigneuric et al., 2000; St Clair-Thompson & Gathercole, 2006). The findings of the current study provide further evidence of these relationships, which of course have important implications for educational practice.

The close relationships between working memory and attainment observed in the current study support previous suggestions that working memory tasks can be used to accurately identify children at risk from poor academic attainment. Furthermore, rather than lengthy comprehensive assessments of working memory, the current study used a brief computerised assessment; Lucid recall (St Clair-Thompson, 2013). Using this assessment it takes approximately 25 minutes to test an entire class of children. Therefore, children could be assessed at any point in childhood using a quick and feasible assessment, which allows educators to predict children's academic attainment. Such assessments could then be used to inform classroom practice and additional educational support. For example, there are many methods for reducing the working memory demands of common classroom activities (e.g. see Alloway & Gathercole, 2006; Gathercole & Alloway, 2008). The results of the current study also suggest that if the goal of a study or assessment is to predict as much variance as possible in children's attainment, including processing speed measures may boost the predictive ability of working memory tasks (see also Cowan et al., 2003). It is also worthy of note that close relationships between working memory and attainment have sparked a great interested in whether adaptive working memory training can enhance working memory capacity and associated higher-order cognitive functioning (e.g. see the final chapters of this book). The relationships between speed of processing and attainment that were observed in the current study suggest that exploring training regimes that target speed of processing may also prove beneficial.

CONCLUSION

The current study revealed close relationships between each component of the multiple-component model of working memory (Baddeley, 2000; Baddeley & Hitch, 1974) and children's reading, writing and mathematics attainment throughout childhood. Thus, working memory is a robust predictor of scholastic attainment with few developmental differences in its predictive ability. The study also revealed that speeds of processing in working memory can predict additional variance in attainment, but only in children aged 11- 13 years. This may reflect the methods used to assess children's learning and attainment. However, further research is needed to further examine relationships between memory storage, processing and attainment, in order to develop a more detailed understanding of the cognitive underpinnings of relationships between working memory and children's learning.

REFERENCES

Alloway, T.P., (2011). A comparison or working memory profiles in children with ADHD and DCD. *Child Neuropsychology, 17,* 483-494.

Alloway, T.P., & Alloway, R.G. (2010). Investigating the predictive roles of working memory and IQ in academic attainment. *Journal of Experimental Child Psychology, 106,* 20-29.

Alloway, T.P., & Gathercole, S. E. (2006). How does working memory work in the classroom? *Educational Research and Review, 1,* 134- 139.

Alloway, T.P., Gathercole, S.E., Willis, C., & Adams, A.M. (2005). Working memory and special educational needs. *Educational and Child Psychology, 22,* 56-67.

Alloway, T.P., & Passolunghi, M.C. (2011). The relationship between working memory, IQ, and mathematical skills in children. *Learning and Individual Differences, 21,* 133-137.

Alt, M .(2011). Phonological working memory impairments in children with specific language impairment: Where does the problem lie? *Journal of Communication Disorders, 44,* 173-185.

Altemeier, L., Jones, J., Abott, R.D., & Berninger, V.W. (2006). Executive functions in becoming writing readers and reading writers: Note taking and report writing in third and fifth graders. *Developmental Neuropsychology, 9,* 71-82.

Archibald, L.M., & Gathercole, S. E. (2006). Short-term and working memory in Specific Language Impairment. *International Journal of Language and Communication Disorders, 41,* 675-693.

Baddeley, A.D. (1986). *Working Memory.* Oxford: Oxford University Press.

Baddeley, A.D. (2000). The episodic buffer: a new component of working memory? *Trends in Cognitive Sciences, 11(4),* 417- 423.

Baddeley, A.D. (2007). *Working memory, thought and action.* Oxford: Oxford University Press.

Baddeley, A.D., & Hitch, G. (1974). Working Memory. In Bower, G.H. (Ed.), *The Psychology of Learning and Motivation.* New York: Academic Press.

Baddeley, A.D., & Logie, R. (1999). Working memory: The multiple component model. In Miyake, A., and Shah, P. (Eds.), *Models of Working Memory* (pp. 28- 61). New York: Cambridge University Press.

Bayliss, D.M., Jarrold, C., Gunn, D.M., & Baddeley, A.D. (2003). The complexities of complex span: Explaining individual differences in working memory in children and adults. *Journal of Experimental Psychology General, 132,* 71-92.

Bayliss, D.M., Jarrold, C., Baddeley, A.D., Gunn, D.M. & Leigh, E. (2005) Mapping the developmental constraints on working memory in span performance. Developmental Psychology, 41, 579-597.

Barrouillet P, Bernardin S, Camos V. (2004). Time constraints and resource sharing in adults' working memory spans. *Journal of Experimental Psychology General, 133,* 83-100.

Bull, R., & Scerif, G. (2001). Executive functioning as a predictor of children's mathematical ability: Inhibition, Switching and Working memory. *Developmental Neuropsychology, 19,* 273- 293.

Cain, K., Oakhill, J., & Bryant, P. (2004). Children's reading comprehension ability: Concurrent predictions by working memory, verbal ability, and component skills. *Journal of Educational Psychology, 96,* 31- 42.

Case, R., Kurland, D.M., & Goldberg, J. (1982). Operational efficiency and the growth of short- term memory span. *Journal of Experimental Child Psychology, 33,* 386- 404.

Conway, A.R.A., Jarrold, C., Kane, M.J., & Towse, J.N. (2007). *Variation in working memory.* NY: Oxford University Press.

Cowan, N., Towse, J.N., Hamilton, Z., Saults, J.S., Elliott, E.M., Lacey, J.F., Moreno, M.V., & Hitch, G.J. (2003). Children's working-memory processes: A response-timing analysis. *Journal of Experimental Psychology: General, 132,* 113-132.

D'Amico, A., & Guarnera, M. (2005). Exploring working memory in children with low arithmetical achievement. *Learning and Individual Differences, 15,* 189-202.

Daneman, M., & Carpenter, P.A. (1980). Individual differences in working memory and reading. *Journal of verbal learning and verbal behaviour, 19,* 450-466.

De Jong, P.P. (1998). Working memory deficits of reading disabled children. *Journal of Experimental Child Psychology, 70(2),* 75- 96.

DeStefano, D., & LeFevre, J. (2004). The role of working memory in mental arithmetic. *European Journal of Cognitive Psychology, 16,* 353- 386.

Duff, S.C., & Logie, R.H. (2001). Processing and storage in working memory span. *The Quarterly Journal of Experimental Psychology, 54,* 31-48.

Engel de Abreu, P.M.J., Gathercole, S.E., & Martin, R. (2011). Disentangling the relationship between working memory and language: The roles of short-term storage and cognitive control. *Learning and Individual Differences, 21,* 569- 574.

Engle, R.W., & Kane, M.J. (2004). Executive attention, working memory capacity, and a two-factor theory of cognitive control. In B. Ross (Ed), *The psychology of learning and motivation* (Vol. 44, pp. 145-199). New York: Elsevier.

Engle, R.W., Kane, M.J., & Tuholski, S.W. (1999). Individual differences in working memory capacity and what they tell us about controlled attention, general fluid intelligence and functions of the prefrontal cortex. In Miyake, A., & Shah, P. (Eds). *Models of working memory: Mechanisms of active maintenance and executive control* (pp. 102-134). London: Cambridge Press.

Engle, R.W., Tuholski, S.W., Laughlin, J.E., & Conway, A.R.A. (1999). Working memory, short-term memory and general fluid intelligence: A latent variable approach. *Journal of Experimental Psychology: General, 128,* 309-331.

Gathercole, S.E., & Alloway, T.P. (2008). *Working memory and learning: A practical guide for teachers.* London: Sage.

Gathercole, S.E., Brown, L., & Pickering, S.J. (2003). Working memory assessments at school entry as longitudinal predictors of National Curriculum attainment levels. *Educational and Child Psychology, 20,* 109- 122.

Gathercole, S.E., & Pickering, S.J. (2000). Working memory deficits in children with low achievement in the national curriculum at 7 year of age. *British Journal of Educational Psychology, 70,* 177-194.

Gathercole, S.E., Pickering, S.J., Knight, C., & Stegmann, Z. (2004). Working memory skills and educational attainment: Evidence from national curriculum assessments at 7 and 14 years of age. *Applied Cognitive Psychology, 18,* 1- 16.

Hitch, G.J., Towse, J.N., & Hutton, U. (2001). What limits children's working memory span? Theoretical accounts and applications for scholastic development. *Journal of Experimental Psychology General, 130,* 184-198.

Holmes, J., Adams, J.W., & Hamilton, C.J. (2008). The relationship between visuo-spatial sketchpad capacity and children's mathematical skills. *European Journal of Cognitive Psychology, 20,* 272- 289.

Imbo, I., Vandierendonck, A. (2007). The role of phonological and executive working memory resources in simple arithmetic strategies. *European Journal of Cognitive Psychology, 19,* 910-933.

Jarrold, C., Baddeley, A.D., Hewes, A.K., Leeke, T.C., & Philips, C.E. (2004). What links verbal short-term memory performance and vocabulary level? Evidence of changing relationships among individuals with learning disability. *Journal of Memory and Language, 50,* 134-148.

Jarvis, H.L., & Gathercole, S.E. (2003). Verbal and nonverbal working memory and achievements on national curriculum tests at 11 and 14 years of age. *Educational and Child Psychology, 20,* 123- 140.

Jeffries, S. & Everatt, J. (2004). Working Memory: its role in dyslexia and other specific learning difficulties. *Dyslexia: an international journal of research and practice, 10,* 196-214.

Kane, M. J., & Engle, R. W. (2003). Working memory capacity and the control of attention: The contributions of goal neglect, response competition, and task set to Stroop interference. *Journal of Experimental Psychology: General, 132,* 47-70.

Kibby, M., Marks, W., Morgan, & S., Long, C. (2004). Specific Impairment in Developmental Reading Disabilities: A Working Memory Approach. *Journal of Learning Disabilities, 37.*

LaPointe, L.B., & Engle, R.W. (1990). Simple and complex word spans as measures of working memory capacity. *Journal of Experimental Psychology: Learning, Memory, and cognition, 16,* 1118-1133.

Majerus, S., Poncelet, M., Greffe, C., & van der Linden, M. (2006). Relations between vocabulary development and verbal short-term memory: The relative importance of short-term memory for serial order and item information. *Journal of Experimental Child Psychology, 93,* 95- 119.

Masoura, E. V., & Gathercole, S. E. (2005). Phonological short-term memory skills and new word learning in young Greek children. *Memory, 13,* 422-429.

Menghini, D., Finzi, A., Carlesimo, G.A., & Vicari, S. (2011). Working memory impairment in children with developmental dyslexia: Is it just a phonological deficit? *Developmental Neuropsychology, 36,* 199-213.

Miyake, A., & Shah, P. (1999). *Models of working memory: Mechanisms of active maintenance and executive control.* New York: Cambridge University Press.

Oberauer, K. (2005). The measurement of working memory capacity. In O.Wilhelm & R. W. Engle (Eds.), Handbook of understanding and measuring intelligence (pp 393-408). Thousand Oaks: Sage.

Pickering, S., & Gathercole, S. (2001). Working Memory Test Battery for Children (WMTB-C). London: The Psychological Corporation.

Raghubar, K.P., Barnes, M.A., & Hecht, S.A. (2010). Working memory and mathematics: A review of developmental, individual difference, and cognitive approaches. *Learning and Individual Differences, 20,* 110-122.

Rourke, B.P. (1993). Arithmetic disabilities, specific and otherwise: A neuropsychological perspective. *Journal of Learning Disabilities, 26,* 214-226.

Seigneuric, A., Ehrlich, M.F., Oakhill, J.V., & Yuill, N.M. (2000). Working memory resources and children's reading comprehension. *Reading and Writing, 13,* 81- 103.

Seitz, K., & Schumann-Hengsteler, R. (2000). Mental multiplication and working memory. *European Journal of Cognitive Psychology, 12,* 552-570.

Seyler, D.L., Kirk, E.P., & Ashcraft, M.H. (2003). Elementary subtraction. *Journal of Experimental Psychology, Learning, Memory and Cognition, 29,* 1339-1352.

Sowerby, P., Seal, S., & Tripp, G. (2011). Working memory deficits in ADHD: The contribution of age, learning/language difficulties and task parameters. *Journal of Attention Disorders, 15,* 461-472.

St Clair-Thompson, H.L. (submitted). Establishing the reliability and validity of a computerised assessment of children's working memory for use in group settings.

St Clair- Thompson, H. L. (2013). *Lucid Recall.* Beverley, East Yorkshire: Lucid Research Limited.

St Clair- Thompson, H.L. & Gathercole, S.E. (2006). Executive Functions and Achievements in School: Shifting, Updating, Inhibition, and Working Memory. *Quarterly Journal of Experimental Psychology A, 59,* 745-759.

St Clair-Thompson, H.L., & Sykes, S. (2010). Scoring methods and the predictive ability of working memory tasks. *Behavior Research Methods, 42, 969-975.*

Swanson, L., & Kim, K. (2007). Working memory, short-term memory, and naming speed as predictors of children's mathematical performance. *Intelligence, 35,* 151-168.

Towse, J.N., Hitch, G.J., & Hutton, U. (2000). On the interpretation of working memory span in adults. *Memory & Cognition, 28,* 341-348

Trbovich, P.L., & LeFevre, J.A. (2003). Phonological and visual working memory in mental addition. *Memory and Cognition, 31,* 738- 745.

Unsworth, N., & Engle, R.W. (2007). On the division of short-term and working memory: An examination of simple and complex span and their relation to higher order abilities. *Psychological Bulletin, 133,* 1038- 1066.

Unsworth, N., Heitz, R.P., Schrock, J.C., & Engle, R.W. (2005). An automated version of the operation span task. *Behavior Research Methods, 37,* 498-505.

Waters, G.S., & Caplan, D. (1996). The measurement of verbal working memory capacity and its relation to reading comprehension. *Quarterly Journal of Experimental Psychology, 49,* 51-79.

Chapter 2

DIFFICULTY OF CHILDREN WITH NORMAL BUT RELATIVELY POOR WORKING MEMORY AT JAPANESE AND MATHEMATICS CLASSES

Masamichi Yuzawa[1] and Miki Yuzawa[2]
[1]Graduate School of Education, Hiroshima University, Hiroshima, Japan
[2]Department of Child Welfare, Faculty of Human Life Sciences,
Notre Dame Seishin University, Okayama, Japan

ABSTRACT

In this chapter, we present research examining whether children with normal but relatively poor working memory experience difficulties in Japanese-language and mathematics classes. Seventy-nine Japanese children (aged 6–7 years) in two first-grade classes were tested for working memory, and six were selected for observation during Japanese and mathematics classes. Although these children had the lowest working-memory scores in each of the two classes, their scores were near the average for their age, as the classes were restricted to children with superior intellectual ability. Although the children nearly always followed their teachers' instructions accurately, they raised their hands relatively rarely when the teachers asked questions. Moreover, they did not pay attention when peers spoke and therefore found it difficult to participate in class discussions among peers. We also analyzed situations in which these children were more likely to raise their hands and identified three ways in which teachers successfully encouraged these children to raise their hands: a) providing sufficient time for the children to think before answering a question, b) repeating questions, and c) offering several alternative answers to a question. We then observed these children in Japanese and mathematics classes 1 year later, when they were in the second grade (aged 8–9 years). The rates at which these students raised their hands in response to teachers' questions or participated in learning activities during Japanese and mathematics classes did not significantly change. These results suggest that children with normal but relatively poor working memories encounter difficulties at school and require appropriate support from teachers.

INTRODUCTION

Working memory, often called a mental workplace, is the system responsible for temporarily storing and manipulating information. Although later revised by Baddeley (2000) to include another element (an episodic buffer), the model proposed by Baddeley and Hitch (1974), which postulated that this system consists of three limited-capacity elements (a central executive, a phonological loop, and a visuospatial sketchpad), has been the basis of much research on working memory. The central executive is a domain-general component responsible for control of the attention and processing involved in a range of regulatory functions, including the retrieval of information from long-term memory. On the other hand, the phonological loop and visuospatial sketchpad provide domain-specific temporary storage for verbal information and visual and spatial representations, respectively. In this chapter, we use the terms "verbal short-term memory" and "visuospatial short-term memory" to refer to the functions of the phonological loop and the visuospatial sketchpad, respectively. We use the terms "verbal working memory" and "visuospatial working memory" to refer to the coordinated function of the phonological loop and the central executive and that of the visuospatial sketchpad and the central executive, respectively. This framework was adopted because we used the Automated Working Memory Assessment (AWMA, Alloway, 2007) to assess working memory in children, and this instrument follows the model developed by Baddeley and Hitch (1974), employing these terms to refer to the functional components of the model (Alloway, 2009; Gathercole & Alloway, 2008).

Working memory plays a key role in supporting children's learning of language and mathematics (Baddeley, Gathercole, Papagno, 1998; Gathercole, 2006; Raghubar, Barnes, & Hecht, 2010; Swanson & Alloway, 2012). Much research has suggested that a poor working-memory capacity is related to learning difficulties and low achievement in areas such as language, mathematics, and science (e.g., Alloway, 2009; Alloway & Alloway, 2010; Gathercole & Alloway, 2008; Dehn, 2008; Pickering, 2006). For example, Alloway, Gathercole, Kirkwood, and Elliott (2009) administered screening tests focused on verbal working memory to 3189 children 5–11 years of age and selected those whose scores were at or below the 10th percentile for children in their age group. These children completed tests of visuospatial working memory, IQ, vocabulary, reading, and mathematics. Their classroom behavior and self-esteem were also assessed. The children were found to have low scores on visuospatial working memory (the mean standard score = 80.39), reading (the mean standard score =78.62), and mathematics (the mean standard score =79.41). The working-memory skills uniquely explained the variance in the reading and mathematics achievement scores when general abilities related to IQ and vocabulary were controlled. Furthermore, teachers typically judged these children to be highly inattentive, highly distractible, unlikely to remember instructions, and unlikely to complete tasks.

Gathercole, Pickering, Knight, and Stegmann (2004) examined the relationship between working-memory skills and performance on national curriculum assessments in English, mathematics, and science in groups of children aged 7 and 14 years. At 7 years of age, achievement levels in English and mathematics were highly correlated with each other, and these levels were significantly correlated with working-memory skills. Thus, children aged 7 years with superior ability in both English and mathematics scored better on working-memory measures than did children with low or average ability. At 14 years of age, correlations among English, mathematics, and science were high, although achievement levels in English and mathematics were less closely linked than they were at 7 years of age. Achievements in mathematics and science were strongly correlated with working-memory scores, with test scores differing significantly between both the low- and average-ability groups and between the average- and high-ability groups. In contrast, school assessments of children's achievement in English were only weakly associated with working-memory scores, and the three ability groups did not differ significantly.

Moreover, Gathercole, Brown, and Pickering (2003) investigated the relationship between measures of working-memory skills at 4 years of age and performance on national curriculum assessments in English and mathematics 2 years later. They found that early working-memory scores were significantly predictive of children's subsequent achievement with respect to literacy of English.

Thus, substantial evidence supports the existence of a close relationship between working-memory skills and learning achievement in areas such as language and mathematics. One explanation for this relationship is that children with poorer skills in this area fail to remember instructions given by teachers, forget what they are currently doing, and fail to fully and accurately complete tasks in class. These classroom behaviors are thought to result in fewer opportunities to learn, and the accumulation of these lost opportunities may lead to low achievement (Gathercole & Alloway, 2008). Due to major individual differences in working-memory capacity (Gathercole & Alloway, 2008), even among children of the same age, we would expect that teachers' utterances to children would be based on the presumably average working-memory capacity of the children in a given class. Under these circumstances, children with poorer working memories would be expected to encounter difficulty understanding long and complex communications by teachers. Indeed, children with a poor working-memory are judged by teachers to be highly inattentive and distractible (Alloway et al., 2009).

In general, the children included in previous studies had the poorest working-memory capacities in their age group. For example, Alloway et al. (2009) selected children whose working-memory scores were at or below the 10th percentile for children in their age group. More than half this group received additional learning support or were monitored by their teachers and were identified as possibly having a developmental disorder. Therefore, it could be that inattentiveness and high distractibility were not caused by the low working-memory skills but were instead mediated by the characteristics of developmental disorders.

To demonstrate that the characteristic classroom behaviors of children with low working-memory skills are caused by a mismatch between children's working-memory skills and teachers' pedagogical approaches, it would be necessary to show that children with normal but relatively poor working-memory skills would encounter the same difficulties encountered by children with extremely poor working-memory skills. Thus, in the following section, we discuss the results reported by Yuzawa et al. (2010a) and Yuzawa, Watanabe, Mizuguchi, Morita, and Yuzawa (in press) involving children with relatively poor but normal working-memory skills who attended classes for children with superior intellectual ability.

Classroom Behaviors of Children with Relatively Poor but Normal Working Memories

Method

Seventy-nine children in two first-grade classes (5 or 6 years of age, 39 boys and 40 girls) participated in this study. All participants attended a school affiliated with a national university in Japan that required students to pass an entrance examination focused on the auditory comprehension of stories; only children with superior intellectual abilities can pass this test.

The children completed a Japanese version of the Automated Working Memory Assessment (AWMA) (Alloway, 2007). We translated an English version of AWMA into Japanese with the permission of the publisher. The AWMA is a computer-based assessment consisting of 12 subtests. Four groups of three subtests measure four components of working-memory skills: verbal short-term memory, visuospatial short-term memory, verbal working memory, and visuospatial working memory. The eight subtests addressing visuospatial short-term memory and visuospatial working memory in the Japanese version are exactly the same as those addressing these domains in the English version. However, all subtests for verbal short-term memory and verbal working memory, with the exception of Counting Recall, were translated and revised. The English figures in Digit Recall and Backward Digit Recall were replaced by Japanese figures. The words used in Word Recall were nouns containing two morae that are supposed to be acquired by the age of 9 years (Sakamoto, 1984). Nonwords in the Nonword Recall were selected from 130 nonwords with two morae and standard word-likeness ratings of 0.5–1.0 (Yuzawa 2010). Some of the sentences presented in the Listening Recall subtest were translated, and others were newly composed so that young Japanese children could easily judge their accuracy. Following the Listening Span Test developed by Osaka (2002), the item to be recalled was the first word in a sentence.

The children completed the AWMA individually during two testing sessions held in a laboratory of the university; each session lasted 45 minutes. The mean scores for the three subtests of the four components were calculated for each child, and the three children in each of the two classes with the lowest total scores on the four components were selected for observation. We referred to the three children in one class as A, B, and C and to the three in the other class as D, E, and F. The mean standard total scores (the standard scores of verbal short-term memory, visuospatial short-term memory, verbal working memory, and visuospatial working memory) for A, B, and C and for D, E, and F were -1.45 (-0.26, -0.07, -2.26, -2.26), -1.79 (-1.17, -0.71, -1.46, -1.54), -2.43 (-0.22, -1.08, -2.73, -2.27), -2.14 (-0.47, -1.64, -2.19, -2.59), -1.63 (-1.22, -0.77, -1.80, -1.45) and, -1.98 (-1.31, -0.95, -1.69, -2.33), respectively, where the Ms and the SDs of the scores of all the children in each class were 0 and 1, respectively. On the other hand, when the scores obtained by these children for visuospatial short-term and working memory were compared with those of English-speaking children in the same age group, the standard scores for A, B, C, D, E, and F were 0.67, 1.87, 0.53, 0.73 -0.20, and 0.07, respectively, for visuospatial short-term memory and -0.20, 0.87, 0.27, 0.33, 0.93, and 0.20, respectively, for visuospatial working memory. Because scores on the four working-memory components are highly correlated with one another, the observed students appear to have had normal working-memory skills for children of their age.

We observed A, B, and C during 11 Japanese-language and eight mathematics classes, and we observed D, E, and F during 10 Japanese-language and eight mathematics classes. We also made additional observations of the target children and six other children with average total scores on working-memory tests in three Japanese and three mathematics classes. One teacher was in charge of all the Japanese classes. Another teacher was in charge of the mathematics classes for A, B, and C, and a third teacher was in charge of the mathematics classes for D, E, and F. The teacher in charge of the Japanese classes and the teacher in charge of the mathematics classes for A, B, and C knew the identities of the targets of the observation and the purposes of the study, whereas the teacher in charge of the mathematics classes for D, E, and F did not know their identities or the purposes of the study.

Observers sat in front of the classroom and recorded the hand-raising and other behaviors of the target children. They also noted when the target children behaved differently from other children. All classes were videotaped from the back of the classroom, and all the teachers' utterances were transcribed. The behaviors of the teachers and target children were coded as follows:

1) Teacher questions and student hand raising: Observers recorded a "Q" whenever children raised their hands, irrespective of whether the teacher explicitly posed a question; an "O" or an "X" was recorded for each non-target and target child, respectively, who raised her/his hand. After the teacher selected a child and that child responded, the same coding procedure was used when other children raised their hands.

2) Participation in the class: Observers classified situations other than questioning (Q) into four categories: "I" indicated that the teacher issued a directive; "B" indicated that the teacher wrote on a blackboard; "E" indicated that the teacher provided an explanation; and "O" indicated that another child was speaking. More specifically, "I" was coded when the teacher directed the whole class to do something (e.g., read a textbook). Observers coded an "O" when they judged that a target followed the instructions, an "X" when they judged that a target did not follow the instructions, and an "A" when they were not able to judge whether each target followed the instructions. In situations coded as "B," observers coded an "O" when a target copied what the teacher wrote on the blackboard in his/her notebook and an "X" when a target did not. "E" was coded when teachers provided an explanation to the whole class about learning material, such as textbooks and worksheets. Observers coded an "O" when they judged that a target listened to the teacher, an "X" when they judged that the target did not, and an "A" when they were unable to judge whether each target listened to the teacher. In situations coded as "O", observers coded an "O" when they judged that a target listened to the speaker, an "X" when they judged that a target did not listen to the speaker, and an "A" when they were unable to judge whether each target listened to the speaker.

Results

We calculated the rates at which each student raised his/her hand by dividing the number of Os by the sum of Os and Xs for each target child in each class. Table 1 shows the mean hand-raising rate for each target child. We also calculated participation rates by dividing the number of Os by the sum of Os and Xs for each target child in each class. Table 2 shows the mean participation rate for each target child in all four types of situations.

Despite differences in the hand-raising and participation rates of each target child between Japanese and mathematics classes, we also detected individual differences in this regard. A and C rarely raised their hands, but followed the teachers' instructions and listened to the teachers and other children. B often raised his/her hand and was attentive. On the other hand, D and E rarely raised their hands and were often inattentive. F raised her/his hand more frequently and was more attentive than were D and E.

Table 1. Mean rates of hand raising and participation by target children

	Class time	Rates of hand rising			Rates of participation		
		A	B	C	A	B	C
Japanese	11	.14	.60	.17	.84	.70	.81
Mathematics	8	.14	.71	.16	.84	.65	.73
		D	E	F	D	E	F
Japanese	10	.18	.17	.31	.39	.48	.64
Mathematics	8	.09	.08	.28	.36	.39	.65

We compared the hand-raising and participation rates among target children with those among children with average total scores on the working-memory tests (Table 2). A, C, D, and E raised their hands at significantly lower rates than did children with average working-memory skills irrespective of the course subject, and all target children participated at lower rates than did children with average working-memory skills irrespective of the subject.

Moreover, we calculated the rates of participation for each target child in each of the four situations in the Japanese and mathematics classes (Table 3). The rates of participation of all target children were higher in situations identified as I and B than in other situations. This suggests that the target children usually followed the teachers' instructions accurately. On the other hand, the rates of participation in O and E situations were generally lower among all target children, which suggests that the target children found it difficult to listen to teachers' explanations of learning material and to comments made by other children.

Table 2. Mean rates of hand raising and participation by target children and children with average working-memory skills

	Class time	Rates of hand rising				Class time	Rates of participation			
		A	B	C	Average		A	B	C	Average
Japanese	1	.21**	.55	.21**	.51	1	.73**	.45*	.85+	1.00
Mathematics	1	.18**	.63	.20**	47	1	.89	.72*	.79	.91
		D	E	F	Average		D	E	F	Average
Japanese	2	.12**	.07**	.29	.29	2	.43**	.35**	.63**	.91
Mathematics	2	.06**	.07**	.35	.29	2	.21**	.55**	.34**	.85

**$p < .01$, *$p < .05$, +$p < .10$: Rates that were significantly lower than those of children with average working-memory skills

Table 3. Rates of participation by target children according to situation

	Instruction			Blackboard			Other			Explanation		
	A	B	C	A	B	C	A	B	C	A	B	C
Japanese	.87	.86	.93	1.00	1.00	1.00	.89	.54**	.79	.75**	.52	.73**
Mathematics	.90	.79	.92	.91	.94	.97	.84	.56**	.65**	.72	.67	.61
	D	E	F	D	E	F	D	E	F	D	E	F
Japanese	.62	.73	.91	1.00	.33	1.00	.14**	.27**	.48**	.50	.33	.38**
Mathematics	.64	.71	.85	1.00	1.00	1.00	.22	.13**	.51**	.13**	.27**	.51

** $p < .01$: Situations in which participation rates were significantly lower

Discussion

The research reported in this section (Yuzawa et al., 2010a, in press) involved observations of the classroom behaviors of children who had the poorest scores on working memory in their classes but whose scores were in the normal range for their age. The results suggest that some of these children rarely raised their hands in class and usually found it difficult to listen to teachers' explanations of learning material and to other children's comments. These behavioral characteristics were consistent with those reported for children with the poorest working memories in their age group (Alloway et al., 2009: Gathercole & Alloway, 2008). Teachers typically judged children with poorest working memory in their age group to be highly inattentive and distractible and to have a short attention span (Alloway et al., 2009; Gathercole & Alloway, 2008). Findings that children with normal but relatively poor working-memory skills showed behavioral characteristics similar to those shown by children with extremely poor working-memory skills suggests that these behaviors may be caused by a mismatch between children's working-memory skills and teachers' pedagogical approaches. As shown in the classes observed for this study, there are large individual differences in the working-memory capacities of children of the same age (Gathercole & Alloway, 2008). The target children were able to correctly follow the teachers' directives because these instructions were generally short and delivered slowly. However, because many children in the classes had superior intellectual abilities, their comments were often lengthy and included complex sentences and ideas. Accordingly, the teachers' explanations also tended to be lengthy and complicated. These lengthy and complicated utterances seemed to exceed the working-memory capacities of the target children, making it difficult for them to listen and join the classroom discussion. These difficulties may lead to low achievement if proper interventions are not provided.

TEACHER SUPPORT FOR CHILDREN WITH POOR WORKING MEMORY

The previous section presented research in which we observed the classroom behaviors of children. The target children obtained the poorest scores in their class on working memory, but their working memory was nonetheless normal for their age. Some of these children rarely raised their hands in class, and these children generally found it difficult to listen to teachers' explanations about learning material and to other children's comments. Because the findings suggested that these outcomes were caused by a mismatch between the children's working-memory skills and the teachers' pedagogical approaches, it would be expected that changes in the latter would lead to changes in the classroom behavior of children with poor working memory. This raises questions about how teachers can support these children. Specifically, could teachers help them to raise their hands and join the class discussion?

To answer these questions, we analyzed those situations in which the target children did not raise their hands initially but did do so eventually (Yuzawa et al., 2010b, in press). Indeed, we identified several strategies used by teachers to stimulate hand raising when only a few children initially responded to their questions. Moreover, even the children with the poorest working memory sometimes responded to these strategies. We identified three strategies associated with hand raising among those children with the poorest working-memory skills who generally rarely raised their hands.

A) Allowing Time for Children to Consider a Question

This strategy involved different approaches to giving children time. Following the first approach, a teacher would ask a question and when only a few children in the class raised their hands, the teacher would give them more time to consider the question. The teacher would tell the children to use the time to reread the material or to talk with other children about the question. Using another approach, the teacher would ask the children to take time to consider a question and, sometimes, to write their ideas in notebooks.

The following is one example drawn from a Japanese class:

Teacher: "It says it can communicate something to others. It can communicate something. What can it communicate?" (Some children raised their hands.)

Teacher: "It can communicate something? There should be something to be communicated. What can it communicate?" (Neither A nor C raised his/her hand.)

Teacher: "All right. *I will give you another chance to read the sentences. You can read it again.* What can it communicate?" (A briefly raised her/his hand but then lowered it and read the textbook again.)

Teacher: "What can it communicate? What does it communicate?"

(A raised his/her hand and was called on.)

B) Repeating a Question

When following this strategy, the teacher would ask a question and, when only a few children in the class raised their hands, he/she would repeat the question.

The following example, drawn from a mathematics class, involved children solving story problems by looking at pictures of monkeys presented on a blackboard.

Teacher: (pointing to the pictures of monkeys) "Consider the number of monkeys. How many monkeys are there?" (Some children raised their hands.)

Teacher: "Look here! *How many monkeys are there?* You all should know this. Someone who is not raising a hand is not looking." (More children raised their hands.)

Teacher: *"How many monkeys are there?"*

C) Asking a Question and Providing Several Alternative Answers

When following this strategy, a teacher would initially asked a question and offer alternative answers from among those previously provided by the children. The teacher would then ask the children to select one answer by raising their hands. This strategy was employed when the teacher judged the question too difficult to answer.

The following example, drawn from Japanese class, illustrates both this strategy and the strategy of allowing time.

Teacher: "This sentence starts with *we* without this phrase. Who are *we*?" (Some children were called on.)
Child 1: "Boys and girls."
Child 2: "All boys and girls. Everyone."
Child 3: "Everything including animals and Americans."
Teacher: "Anyone have a different idea?"
Child 4: "People."
Child 5 "People in the world."
Child 6: "*I* means the self, but *we* means everyone."
Child 7: "All the creatures including animals."
(Children talked with each other in pairs.)
Teacher: *"Please choose one. Those who think we means people? Those who thinks we means all the creatures including people and other animals? Those who are between the two, who think we means people and some animals?"*
(D and E raised their hands in response to one of the alternatives.)

We counted all the instances in which these three strategies were used in the 11 Japanese and eight mathematics classes where A and C participated and in the 10 Japanese and eight mathematics classes where D and E participated. We then calculated the rates at which these children raised their hands in each situation (Table 4). The rates at which they raised their hands were significantly higher in these than in other situations where teachers posed questions. These data suggest that the three strategies were effective in helping children with poor working memory to raise their hands and participate in class discussions.

Table 4. Rates of hand raising by A, C, D, and E according to situation

	Number[1]	A	C	Number[2]	D	E
Giving Time	13	.54**	.69**	13	.46*	.23**
Question Repetition	11	.27*	.36+	5	.20	.20
Giving Alternatives	7	1.00**	1.00**	11	.82**	.55**

Note: [1]Number of situations in the 11 Japanese and eight mathematics classes, [2]Number of situations in the 10 Japanese and eight mathematics classes.
+$p<.10$, *$p<.05$, **$p<.01$: significantly higher than in other situations involving questions

Changing the Classroom Behaviors of Children with Relatively Poor but Normal Working Memory Memories

In the previous sections, we reported on research (Yuzawa et al., 2010a, 2010b, in press) in which we observed the classroom behavior of children with working-memory scores that were in the normal range for their age but were the lowest in their class. Some of these children rarely raised their hands during class, and they generally found it difficult to listen to teachers' explanations about learning material and to other children's comments. Our analysis of situations in which these children did not raise their hands initially but did do so later revealed three strategies that were effective in helping these children raise their hands and participate in class discussions.

Findings showing that children with normal but relatively poor working-memory skills encountered difficulties similar to those faced by children with extremely poor working-memory skills suggest that the characteristic classroom behaviors of children with poor working memory are caused by a mismatch between the children's working-memory skills and the teachers' pedagogical approaches. Thus, we posited that these difficulties may be overcome by altering the methods of instruction. Indeed, several strategies were effective in helping children increase their participation, inviting the question of whether these children would change their classroom behavior in the future.

It was possible that once teachers were able to identify the children with the poorest working memories, who therefore were likely to struggle during class, they would pay attention to these children and support their participation in classroom discussions. Such support, in turn, may promote participation. Indeed, the teachers did intentionally devote attention to these children and often provided support to them.

It was also possible that, despite this support from teachers, the children's classroom behavior would change only minimally, given the difficulty involved in changing children's verbal behavior during class discussions. Elementary school teachers in Japan often emphasize class discussions; in such situations, learning involves communications between the teacher and the children and among the children themselves. The classes we observed included many children with superior intellectual abilities, and these children actively participated in class discussions by contributing lengthy and complex comments expressing complicated ideas. Accordingly, the teachers' responses to these ideas were also complicated. As noted above, children with the poorest working memories in the class found it difficult to listen to other children's comments and to utterances made by teachers. This situation would not be expected to change in the future, and children with the poorest working memories in the class would be likely to continue encountering difficulties.

To test these possibilities, we observed the same target children 1 year later, when they were in the second grade. We observed A, B, and C during five Japanese-language and three mathematics classes and D, E, and F during six Japanese-language and three mathematics classes. The teachers of these classes were the same as those in the first grade. The methods of observation were the same as those outlined in the second section of this chapter.

Table 5 shows the mean hand-raising and participation rates by the target children during Japanese and mathematics classes. Tables 1 and 5 show that the mean hand-raising and participation rates were surprisingly consistent across time. One year later, A and C rarely raised their hands but did follow teachers' instructions and listened to teachers and other children. B often raised her/his hand and was attentive. On the other hand, D and E rarely raised their hands and were inattentive. F raised her/his hands more frequently and was more attentive than were D and E. We found similar behavioral characteristics in the Japanese and mathematics classes.

These results support the second possibility noted above, namely that the classroom behavior of these children would change only minimally despite support from teachers. These results were disappointing, especially to the teachers, who had intentionally paid attention to the target children and provided prompts to encourage their participation in classroom activities over the course of the previous year. In the same vein, Elliot, Gathercole, Alloway, Holmes, and Kirkwood (2010) implemented a classroom-based intervention to help primary school children identified as having working-memory difficulties. The intervention involved providing the teaching staff with information about working memory and the problems that children with poor working memories encounter; they were also shown how they might adjust their teaching so that the memory-load demands would be appropriate for this group of students. Additionally, they received training in how to encourage these children to use strategies to minimize memory load. However, no evidence that the intervention program had an effect on the children's working memory or academic performance was found.

Given the major individual differences in the working memory of children in a single class, the disadvantages faced by children with poorer working memories would be expected to be relatively enduring, underscoring the need for teachers to persevere in their efforts to help this group. One promising finding of our study (Yuzawa et al., 2011) was that teachers used revoicing (O'Connor & Michaels, 1996) during classes in which the children with the poorest working memories had higher rates of participation; that is, the teacher rephrased and summarized children's lengthy, complex, or vague comments into short and concise statements, added appropriate explanations, and organized the content. This approach enabled other children to understand the comments and join the discussion more easily. It was suggested that the use of such revoicing by teachers could be helpful to children with the poorest working memories.

Table 5. Mean rates of hand raising and participation by the target children

	Class time	Rates of hand rising			Rates of participation		
		A	B	C	A	B	C
Japanese	5	.13	.59	.33	.82	.71	.80
Mathematics	3	.07	.49	.10	.72	.56	.71
		D	E	F	D	E	F
Japanese	6	.15	.08	.18	.48	.39	.64
Mathematics	3	.06	.04	.18	.26	.51	.41

CONCLUSION

In this chapter, we presented our research examining whether children with normal but relatively poor working memories encountered difficulties during Japanese-language and mathematics classes. Children with the poorest working-memory scores in classes for children with superior intellectual abilities were selected for observation. These children nearly always followed teachers' instructions accurately, but they raised their hands relatively rarely in response to teachers' questions and did not pay attention when peers spoke. Therefore, they found it difficult to participate in class discussions involving peers. The behavioral characteristics of these children were similar to those that have been reported among children with extremely poor working-memory skills (e.g., Alloway et al., 2009; Gathercole & Alloway, 2008). These finding suggest that these behaviors may be caused by a mismatch between children's working-memory skills and teachers' pedagogical approaches. Children's comments and teachers' explanations tended to be lengthy and complex, and it seemed that the children with the poorest working memories found it difficult to follow and understand these comments and explanations.

Our analysis of situations in which these children raised their hands after initially declining to do so identified three ways in which teachers successfully encouraged this group to raise their hands: a) allowing enough time for children to consider a question, b) repeating questions, and c) offering several alternatives as answers to a question. Teachers' intentional use of these approaches may be effective in stimulating these children's participation in learning activities. However, our observations of these children during Japanese and mathematics classes 1 year later, when they were in the second grade (aged 8–9 years), revealed that the rates at which these students raised their hands in response to teachers' questions and participated in learning activities during Japanese and mathematics classes had changed only minimally. These results suggest that children with poorer working memories encounter relatively enduring difficulties and need appropriate support due to the major individual differences in working-memory capacities among children in the same class. The kinds of support that would be useful should be explored in future research.

ACKNOWLEDGMENTS

Note: The preparation of this chapter was supported in part by the Grant-in-Aid for Scientific Research (B), 23330203.

REFERENCES

Alloway, T.P. (2007). *Automated Working Memory Assessment*. London: Psychological Corporation.
Alloway, T.P. (2009). *Improving working memory: Supporting students' learning*. London: Sage.
Alloway, T. P., & Alloway, R. G. (2010). Investigating the predictive roles of working memory and IQ in academic attainment. *Journal of Experimental Child Psychology, 106*, 20-29.
Alloway, T.P., Gathercole, S.E., Kirkwood, H., & Elliott, J. (2009). The cognitive and behavioral characteristics of children with low working memory. *Child Development, 80*, 606-621.

Baddeley, A. D. (2000). The episodic buffer: A new component of working memory? *Trends in Cognitive Sciences, 4*, 417-423.

Baddeley A. D., Gathercole, S.E., & Papagno, C. (1998). The phonological loop as a language device. *Psychological Review, 105*, 158-173.

Baddeley, A.D. & Hitch, G. (1974). Working memory. In G. A. Bower (Ed.), *The psychology of learning and motivation, 8*, 47-90, New York: Academic Press.

Dehn, M. J. (2008). *Working memory and academic learning: Assessment and intervention.* Hoboken, NJ: Wiley.

Elliot, J. G., Gathercole, S. E., Alloway, T. P., Holmes, J., & Kirkwood, H. (2010). An evaluation of a classroom-based intervention to help overcome working memory difficulties and improve long-term academic achievement. *Journal of Cognitive Education and Psychology, 9*, 227-250.

Gathercole, S. E. (2006). Nonword repetition and word learning: The nature of the relationship. *Applied Psycholinguistics, 27*, 513-543.

Gathercole, S. E., & Alloway, T, P. (2008). *Working memory and learning: A practical guide for teachers.* London: Sage.

Gathercole, S. E., Brown, L., & Pickering, S. J. (2003). Working memory assessments at school entry as longitudinal predictors of National Curriculum Attainment levels. *Educational and Child Psychology, 20*, 109-122.

Gathercole, S. E., Pickering, S. J., Knight, C. & Stegmann, Z. (2004). Working memory skills and educational attainment: Evidence from national curriculum assessments at 7 and 14 years of age. *Applied Cognitive Psychology, 18*, 1–16

O'Connor, M. C., & Michaels, S. (1996). Shifting participant frameworks: Orchestrating thinking practices in group discussion. In D. Hicks (Ed.), *Discourse, learning, and schooling.* Cambridge, UK: Cambridge University Press. Pp. 63-103.

Osaka M. (2002). Working memory: Note of the barain. Tokyo: Shinyosya. (In Japanese)

Pickering, S. J. (ed.) (2006). *Working memory and education.* Burlington, MA: Academic Press.

Raghubar, K. P., Barnes. M. A., & Hecht, S. A. (2010). Working memory and mathematics: A review of developmental, individual difference, and cognitive approaches. *Learning and Individual Differences, 20*, 110–122.

Sakamoto, I. (1984). New basic vocabulary for education. Tokyo: Gakugeitosyo. (In Japanese)

Swanson,H. L., & Alloway, T. P. (2012). Working memory , learning, and academic achievement. In K. R. Harris, S. Graham, & T. Urdan, (Editor-in-Chiefs), C. B. McCormick, G. M. Sinatra, & J. Sweller, (Associate Eds.), *APA educational psychology handbook. Vol. 1, Theories, constructs, and critical issues.* Washington, DC: APA.

Yuzawa, M. (2010). Studies about phonological short-term memory in young children. Tokyo: Kazama-syobo. (In Japanese)

Yuzawa, M., Aoyama, Y., Ito, K., Maeda, K., Nakata, S., Miyatani, M., Chujo, K., Sugimura, S., Morita, A., Yamada, K., Kondo, A., Tateishi, Y., Kishita, M., & Mito Y. (2010a). Learning supports for children with poor working memory: Analyses of classroom behavior of first graders with relatively poor working memory. *The Annals of Educational Research, 39*, 39-44. (In Japanese)

Yuzawa, M., Maeda, K., Miyatani, M., Chujo, K., Sugimura, S., Morita, A., Yamada, K., Kondo, (2010b). Development of a data base for special needs based on working memory profiles of individual children. *Reports of Collaborative Research by Graduate School of Education, Hiroshima University, 9*, 105-122. (In Japanese)

Yuzawa, M., Watanabe, D., Maeda, K., Miyatani, M., Chujo, K., Sugimura, S., Morita, A., & Kondo, (2011). Development of a data base for special needs based on working memory profiles of individual children II. *Reports of Collaborative Research by Graduate School of Education, Hiroshima University, 10*, 129-141. (In Japanese)

Yuzawa, M., Watanabe, D., Keigo, M., Morita, A., & Yuzawa, M. (in press). Classroom behavior of children with poor working memory at class and learning supports for them. *The Japanese Journal of Developmental Psychology.* (In Japanese with an English abstract).

In: Working Memory
Editor: Helen St. Clair-Thompson

ISBN: 978-1-62618-927-0
© 2013 Nova Science Publishers, Inc.

Chapter 3

THE ROLE OF WORKING MEMORY AND INHIBITION IN THE DEVELOPMENT OF MOTOR PLANNING BETWEEN 7 AND 10 YEARS OF AGE

Valérie Pennequin
Université François Rabelais, Tours, France

ABSTRACT

This chapter tested to what extent working memory and inhibition are linked to the development of planning in children. In a previous study with children aged 4 to 7 years, we observed correlations between an egg-planning task (Wellman, Sophian & Fabricius, 1985), the inhibition of a pre-potent but inappropriate response, and the manipulation of working memory strategies (Pennequin, Sorel, & Fontaine, 2010). In that study we found a floor effect for children aged over 6 years. In the present study, we adopted a different methodology to test the effect of inhibition and memory load on the planning performance of children aged over 6 years. We manipulated working memory load and inhibition within the task: working memory load was manipulated by increasing the number of sub-goals needed to plan a route, and inhibition by manipulating the pre-potency of an automatic strategy, which may be irrelevant under certain conditions. Sixty-eight children were divided into two groups: group 1 was composed of 31 children aged 7 years, and group 2 was composed of 37 children aged 10 years. Results showed a developmental effect, with planning performance improving between 7 and 10 years. Analyses of variance showed a main effect of memory load: performance deteriorated when the number of sub-goals increased, but in a similar way in the two age groups. An interaction between inhibition and memory load was also observed. In the discussion, we consider how automatic heuristics could enhance or hinder planning performance by acting on memory load. We also discuss the results in relation to the development of each process and the way they interact.

INTRODUCTION

During development, there are significant changes in the ability to control deliberate thoughts and actions (for a review, see Diamond, 2002). These changes have been associated with the development of executive functioning, which is an umbrella term for various cognitive processes that sub-serve goal-directed behaviour (Miller & Cohen, 2001). The executive process that guides responses and enables successful goal-directed behaviour is known as response planning. In everything we do, we are constantly faced with a variety of alternative responses, giving us cognitive control of our behaviour. Planning is an essential part of goal-setting, because it involves the ability to plan actions in advance and to approach a task in an organized, strategic, and efficient manner (Anderson, 2002). In a previous study (Pennequin, Sorel & Fontaine, 2010), we examined the assumption that the executive components of working memory, shifting and inhibition (Miyake, Friedman, Emerson, Witzki, Howerter & Wager, 2000) contribute to motor planning performance and that this contribution changes between 4 and 7 years of age. The planning task used in our previous study (Pennequin et al., 2010) was based on an egg-planning task (Wellman, Sophian & Fabricius, 1985) in which the children had to fetch eggs from two different places and put them in a basket which they were not allowed to move. The conditions were varied by putting the eggs and basket in different places. To achieve the task efficiently, the children needed to plan their moves. For example, the most efficient strategy was to go first to the egg which was furthest away and then to the egg nearest the basket. Our results showed that the age effect is complex: planning performance did not improve between the ages of 4 and 7, although the three lower-level executive functions did. In addition, results indicated that inhibition and working memory scores were correlated with planning performance but revealed no evidence for shifting. We concluded that planning capacities do not depend directly on age, but are mediated by the development of lower-level executive functions. For example, the planning capacity of a 7-year-old child with inefficient lower-level executive functions may be inferior to that of a 4-year-old with more efficient working memory and inhibitory functioning. In the literature, the role of inhibition in planning tasks seems to differ according to the participants' age. For example, Miyake et al. (2000) and Huizinga, Dolan and Van der Molen (2006) found that inhibition (measured using the Stroop task) played a role in the planning performance of young adults in the Tower of Hanoï or Tower of London task. However, Huizinga et al. (2006) found that inhibition did *not* predict Tower of London performance in their three child groups (7, 11 and 15 years of age).

The aim of the present study was thus to examine further the development of the relationship between inhibition, working memory and planning, and more specifically to investigate how inhibition and working memory interact to produce goal-oriented behaviour in children over 6 years of age. In the literature dealing with executive functions, inhibition has been found to be related to working memory (Hasher & Zacks, 1988; Hasher, Zacks & May, 1999; Miyake et al. 2000). The findings of Houghton and Tipper (1994) support the assumption that the strength of the inhibition must be continuously adjusted to the strength of the prepotent response that has to be inhibited. The efficiency of inhibitory processes therefore depends on the available resources in working memory. According to Roberts, Hager and Heron (1994), working memory and inhibition are linked in their functioning because participants must keep in mind instructions and relevant information for the ongoing task while simultaneously inhibiting inappropriate prepotent responses. In three studies with adults using a dual-task paradigm, the authors examined the hypothesis that taxing working memory beyond a certain threshold can result in decreased inhibition. Their results indicated that conditions with the highest working-memory load produced inhibitory errors comparable to those made by patients with prefrontal dysfunctions. They hypothesized that task performance depends on the strength of the prepotency of the response to be inhibited, the functioning of working memory, and the memory load of the task. These three characteristics are considered independent because a task could require the inhibition of a very prepotent response but not have a great memory load, as in the antisaccade task or Stroop task. Conversely, a task could place great demands on working memory but not require inhibition of a previous prepotent response, as in the Wisconsin card-sorting task. Kane, Bleckley, Conway and Engle (2001) found that adults with low memory span were slower and less accurate than high-span participants on an antisaccade task. Similar results have been found with the Stroop task (Kane & Engle, 2003), the dichotic listening task (Conway, Cowan & Bunting, 2001), the directed forgetting paradigm (Aslan, Zellner & Bäuml, 2010), and the Go/No-Go task (Redick, Calvo, Gay & Engle, 2011). Negative effects of working memory load on inhibitory control have also been found in other paradigms such as visual selective attention (De Fockert, Rees, Frith & Lavie, 2001; Lavie, Hirst, de Fockert & Viding, 2004) and the Go/No-Go task (Hester & Garavan, 2005; Redick et al., 2011). In sum, experimental data support the idea of a functional link between working memory capacity and inhibitory efficiency in adults and are in accordance with our previous results with children (Pennequin et al., 2010).

The aim of the present study was to extend our previous experiment (Pennequin et al., 2010) concerning the development of planning and its relationship with executive functions. More precisely, we studied two age groups to explore the effect of development on the relationship between working memory load and inhibition in planning ability. The disadvantage of the correlation methodology used in our previous study is that the links between executive components are often task-dependent. In the present study, we used a different methodology, manipulating working memory load and inhibition within the task: working memory load was manipulated by increasing the number of sub-goals, and inhibition by manipulating the pre-potency of an automatic strategy that may be irrelevant under certain conditions. In accordance with the literature, we hypothesized that the efficiency of the inhibition of a prepotent response would depend on the working memory load, particularly for the younger children, and consequently that the development of the ability to plan would depend on the functional link between inhibition and working memory.

METHOD

Participants

The participants were sixty-eight healthy children aged 6 to 10 years. They were divided into two groups: group 1 was composed of 31 children with an average age of 7 years (18 males, $M = 7.2$ years, S.D. = 0.7), and group 2 was composed of 37 children with an average age of 10 years (16 males, $M = 10$ years, S.D. = 0.6). Children were recruited from local schools around the city of Tours in France and were all native French speakers. Teachers assisted in the selection process in order to exclude children with health or behaviour problems.

Measures

The planning task was adapted from Wellman, Sophian and Fabricius (1985). Participants had to imagine that they had to collect two eggs (X and Y) from two different places in the classroom and put them in a basket, without moving the basket. The children were given a picture showing where the eggs were and had to draw on it the most efficient route (fast and without round trips). This involved planning their moves to be as efficient as possible. Inhibition was manipulated by changing the distance of egg X from the child; placing the egg near the child would trigger an automatic strategy, the proximity heuristic, creating a tendency to go first to the nearest goal, even if this was not the most efficient strategy. There were four conditions (see Figure 1):

- In condition A, 'planning', the basket was near egg Y which was furthest from the child. The most efficient strategy was to pick up egg X first, then go to egg Y and put the two eggs in the basket. Egg X was not sufficiently near the child to induce the proximity heuristic.
- In condition B, 'proximity', the basket and egg X were near the child. It was equally efficient to pick up egg X or egg Y first. This condition was designed to indicate whether the child was aware of the proximity heuristic. This awareness was a requirement of the child's participation in the study, because if the child did not use this heuristic automatically, it would not be possible to study the effect of inhibiting this strategy on planning.
- In condition C, 'planning and proximity', the proximity heuristic facilitated planning and corresponded to the most efficient route. In this condition, the basket was near egg Y which was furthest from the child. The most efficient route was therefore to go first to egg X and then to egg Y and put the two eggs in the basket. Comparing conditions A and C indicated whether the proximity heuristic facilitated planning.
- In condition D, 'planning versus proximity', the proximity heuristic had to be inhibited to plan the most efficient route. Comparing conditions D and A showed the interaction of the inhibitory process with planning.

The original task involved only two eggs; this is the lightest memory load level. In the present study, we manipulated working memory load by increasing the number of eggs from two to five, thereby increasing the number of sub-goals involved in planning the best route. In this way, four memory-load levels were constructed corresponding to the number of sub-goals involved. In all, there were 16 trials.

Three quantitative dependent variables were taken into account:

- Planning performance. Performance was scored 0 – 2, with 2 corresponding to optimum planning strategy + proximity strategy

- Planning time, measured as the number of seconds between the end of the instructions and when the child began to draw the route.
- Execution time, measured as the number of seconds taken to draw the route.

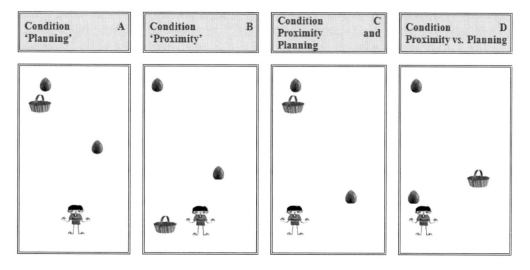

Figure 1. The four conditions for the egg-planning task at level 1 (with two eggs).

Procedure

All the participants were tested individually in a quiet classroom. Each participant was seen only once for a maximum of 30 min. Care was taken to ensure that all the children understood the instructions. The order of the four tasks was counterbalanced between children.

RESULTS

ANOVAs were performed on the three dependent variables: planning performance, planning time and execution time. There were three independent variables: age (with two modalities: 7 and 10 years), condition (with three modalities: A, C, D), and memory load (with four modalities: levels 1, 2, 3, and 4). Condition B (proximity) was excluded from the ANOVA because it was an inclusion requirement, as the inhibition effect could only be studied if the child used the proximity heuristic and inhibited it in condition D. Two children were excluded from the sample (one aged 7 and one aged 10) because they did not use this heuristic at any level. For the statistical analyses, there was thus a group of 30 seven-year-old children and a group of 36 ten-year-old children. Conditions and memory load were repeated measures. When significant effects were observed, planned comparisons were performed for more in-depth analysis of the results.

ANOVA on Planning Performance

For both age groups, a floor effect was observed for condition A (planning) and condition C (planning and proximity), with no variance at level 1 (with two eggs). We therefore removed this level from the ANOVA on planning performance. The four levels were included in the ANOVAs on the other two dependent variables (planning time and execution time).

The ANOVA showed a main effect of age on performance ($F(1, 64) = 4.8$, $p < .05$): 10-year-old children performed better than the 7-year-olds. We also observed a main effect of the number of eggs ($F(2, 128) = 33.31$, $p<.001$); performance decreased when the number of eggs increased from 3 to 5. Performance differed significantly between levels ($F(1, 64) = 19.53$, $p<.001$ between levels 2 and 3; $F(1, 64) = 14.7$, $p<.001$ between levels 3 and 4; $F(1, 64) = 64.59$, $p<.001$ between levels 2 and 4). In other words, children performed worse when they had to plan with five than with four or three eggs. There was also a main effect of condition ($F(3, 198)=15,72$, $p<.001$), with significant differences between performance in condition A and condition C ($F(1, 64) = 27.64$, $p<.001$) and condition D ($F(1, 64) = 14.53$, $p < .001$). In other words, children made more errors in the simple planning condition than in conditions where the proximity heuristic was prepotent. There was no difference between conditions C and D ($F(1, 64) = 0.07$, $p>.05$), children performing equally well in condition D where the proximity heuristic had to be inhibited as in condition C where the proximity heuristic was relevant for planning.

Only the interaction between condition and the number of eggs was significant ($F(4, 256) = 12.33$, $p<.001$) (see Figure 2). The effect of the number of eggs on planning differed according to condition; performance declined as the number of sub-goals increased, but only in conditions A and D. The triple interaction between the three independent variables was not significant ($F(4, 256) = 1.41$, $p>.05$). The interactions between age and condition ($F(2, 128) = 0.76$, $p>.05$) and between age and number of eggs ($F(2, 128) = 0.63$, $p>.05$) were also non-significant.

Another ANOVA was performed on planning time, measured as the time between the end of the instructions and when the child started to draw the route. Level 1 with two eggs was included in this ANOVA because there was no floor effect with response time. For the same reason, condition B (proximity strategy) was also included in this analysis. There were three independent factors: age (7 and 10 years of age), condition (A, B, C, D), and level (2, 3, 4, or 5 eggs). For the sake of clarity, for planned comparisons only significant results have been given.

This ANOVA on planning time showed no main effect of age ($F(1, 66) = 2.45$, $p > .05$), but we did observe a main effect of the number of eggs ($F(3, 198) = 3.09$, $p<.05$). Planned comparisons showed a longer planning time at level 2 (with three eggs) than at the other levels (one, four and five eggs) ($F(1, 66) = 7.39$, $p>.01$ between levels 1 and 2; $F(1, 66) = 8.03$, $p<.01$ between levels 2 and 3; $F(1, 66) = 0.15$, $p>.05$ between levels 3 and 4; $F(1, 66) = .016$, $p>.05$ between levels 1 and 3; $F(1, 66) = .011$, $p>.05$ between levels 1 and 4). There was also a significant effect of condition ($F(3, 198) = 8.51$, $p<.001$). Planning time in condition A was significantly longer than in condition C ($F(1, 66) = 20.11$, $p<.001$) and condition D ($F(1, 66) = 21.6$, $p < .001$).

In other words, children took longer to plan in the simple condition than in conditions where the proximity heuristic was relevant. There was no difference between conditions C and D ($F(1, 66) = 0.03$, $p>.05$), children spending as long to plan in condition D, where the proximity heuristic had to be inhibited, as in condition C where the proximity heuristic was relevant.

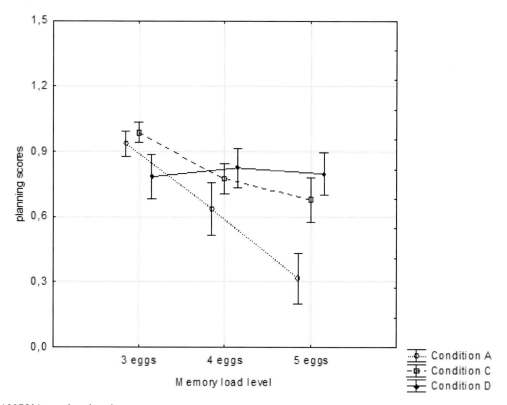

ANOVA on planning time.

Figure 2. Planning performance according to number of eggs and condition.

Like the results for planning performance, the only significant interaction was between condition and number of eggs (F(9, 594) = 4.93, p<.001). The effect of the number of eggs differed according to the condition; the more the eggs, the longer the children took to draw a route, particularly in condition A (F(1, 66) = 9.86, p < .01 between levels 1 and 3; F(1, 66) = 9.23, p < .01 between levels 2 and 3) and condition C (F(1, 66) = 6.49, p = .01 between levels 1 and 2; F(1, 66) = 5.37, p < .05 between levels 3 and 4). For conditions B and D, planning time also decreased when there were more eggs, but less noticeably (F(1, 66) = 6.23, p=.01 between levels 1 and 4 in condition B, and F(1, 66) = 9.72, p<.01 between levels 1 and 2 in condition D). The triple interaction between the three independent variables was not significant (F(9, 594) = 1.49, p>.05). The interactions between age and condition (F(3, 198) = 0.97, p>.05) and between age and number of eggs (F(3, 198) = 0.89, p>.05) were also non-significant.

ANOVA on Execution Time

An ANOVA was performed on the dependent variable of 'execution time', that is, the time children took to respond in each condition and at each level (number of eggs). There were also three independent factors: age, condition and level.

Results were similar to those obtained with the two other dependent variables (performance and planning time). We observed a main effect of age ($F(1, 66) = 14.61$, $p<.001$): the 10-year-old children took longer to execute their plan than the 7-year-olds. A main effect of the number of eggs ($F(3, 198) = 12.33$, $p<.001$) was observed: as the number of eggs increased (from two to five), execution time decreased. We also observed a main effect of condition ($F(3, 198) = 17.05$, $p<.001$). Planned comparisons showed that children took longer to respond in condition D than in condition A ($F(1, 66) = 23.89$, $p<.001$) and condition C ($F(1, 66) = 15.63$, $p<.001$). In other words, children needed more time to execute a plan requiring the inhibition of an irrelevant heuristic.

Like the results for the dependent variables of planning performance and planning time, only the interaction between condition and number of eggs was significant ($F(9, 594) = 5.13$, $p<.001$). The effect of the number of eggs on execution time varied according to the condition: the more eggs there were, the less time children took to draw a route, particularly in condition A ($F(1, 66) = 29.64$, $p<.001$ between levels 1 and 3; $F(1, 66) = 28.24$, $p<.01$ between levels 2 and 3; $F(1, 66) = 6.13$, $p=.01$ between levels 3 and 4). In the other three conditions, the number of eggs did not affect the time taken to draw the route. The triple interaction between the three independent variables was not significant ($F(9, 594) = 1.36$, $p>.05$). The interactions between age and condition ($F(3, 198) = 0.4$, $p>.05$) and between age and number of eggs ($F(3, 198) = 1.56$, $p>.05$) were also non-significant.

To summarize, the three ANOVA results on performance, planning time and execution time were very similar, with a main effect of each independent factor (apart from no age effect on planning time) and an interaction between condition and number of eggs for each dependent measure.

DISCUSSION

The issue addressed by the present study concerns the development of planning and the relationships between working memory load and inhibition in planning ability. Using the egg-planning task, we manipulated memory load by increasing the number of sub-goals (eggs) needed to plan a route; inhibition was manipulated through the pre-potency of an automatic strategy (the proximity heuristic), which was irrelevant under certain conditions. Two age groups were studied, one with a mean age of seven years and the other with a mean age of ten years.

Results showed that the 10-year-old children performed better than the younger group. They also took longer to perform the motor action (drawing the route), suggesting that they developed their plan while carrying out the action rather than before it. No interaction between age and the other independent factors was observed. Thus, we found no evidence of a developmental difference regarding the effects of condition and number of eggs involved in the planning task.

Condition A, which involved no proximity strategy, yielded the worst performance, longest planning time, and shortest execution time. Children developed their plan before executing the motor response, but this planning strategy did not seem to be very efficient. It is possible that planning before beginning the motor response poses a considerable load on working memory, as the instructions and sub-goals (memory span) must be stored in memory, the information needed to plan a route must be manipulated (working memory component), and task-related information must be up-dated while carrying out the motor response. The task calls on all working memory components. Although planning before beginning the motor response would appear to be an efficient strategy, it is possible that it leads children under 10 to make more errors because of the limitations of their working memory abilities.

Under the conditions with the proximity heuristic, children spent less time planning before acting. The proximity of an egg led to an automatic motor response, and the children had difficulty inhibiting their tendency to go into action. In condition D, which required inhibition of the proximity heuristic, execution time was longer, suggesting that the children planned their route from one sub-goal (egg) to the next as they carried out the motor action. This strategy of planning while carrying out the motor response was more efficient for children in both age groups and resulted in better performance. This tendency did not differ between 7 and 10 years of age.

Generally, as the number of sub-goals (eggs) increased, planning performance decreased; planning time also decreased, but only in conditions C and D. In condition A, it was execution time that decreased, and the children who planned before beginning the motor response assigned even less time to executing their plan when the number of eggs increased. By contrast, when the number of sub-goals increased in conditions C and D (with the proximity heuristic), the children who planned their moves while carrying out the motor response assigned even less time to planning before beginning the motor response. In sum, the effect of the number of eggs is complex: in general, the total time taken to solve the task decreased as the number of eggs increased. The relative speed at which children carried out the different stages of the task depended on whether or not the proximity heuristic was present. More precisely, children used their favourite planning strategy (before or during the motor response) when the number of eggs increased and spent less time on the other steps of the task. These results could be interpreted as a tendency to act when the visual salience of the eggs increased. A large number of eggs triggered a bottom-up treatment of the stimuli, in other words an automatic tendency to rush into action. The proximity heuristic and the visual salience of eggs seem to enhance this tendency. Thus, when the number of eggs increased, the children maintained their planning strategy but less effectively, as their performance deteriorated.

We expected that increasing the number of eggs would increase the number of sub-goals and hence the working memory load. However, it seems to be more complicated than that, as increasing the number of eggs also increased their visual salience. This seemed to activate a bottom-up treatment, reducing planning time and performance. In condition D (with the proximity heuristic) with a large number of eggs, the strength of the prepotency of the response to be inhibited appeared to be greater because the motor response was strongly influenced by the proximity of an egg and by the visual salience of the other eggs. Consequently, both execution time and performance decreased. More eggs seemed to increase the inhibition demand and not just the working memory load as we had supposed. These results are in line with the conclusions drawn by Roberts, Hager and Heron (1994), that working memory and inhibition are linked in their functioning because the participants must keep in mind the instructions and relevant information for the task while inhibiting an inappropriate prepotent response. The authors assumed that task performance depends on the strength of the prepotency of the response to be inhibited, on the functioning of working memory, and on the memory load of the task. Our results indicate that the results of that study with adults are also valid with 7- and 10-year–old children.

The importance of considering the three measures - planning performance, planning time and motor execution time – is clear. If we had measured only performance, we would have concluded wrongly that the poorer performance with a greater number of eggs was due to an increase in the number of sub-goals that children had to retain in working memory. In fact, when there were many eggs and one near them, the visual salience of the sub-goals was greater and the children preferred to plan while plotting the route rather than trying to keep all the sub-goals in working memory before beginning to draw the route. Their planning strategy required keeping only the next sub-goal (the next egg) in working memory and updating sub-goals as they went on. Thus, they used the executive component of working memory (updating) and relied less on the storage component (memory span). Despite the poorer performance when there were more sub-goals, it seems that the children adopted a planning strategy that corresponded to their working memory capacities. We therefore hypothesize that the poorer performance when the number of sub-goals increased is not necessarily an indication of inefficient cognitive processes, but could reflect a shifting ability based on the task demands.

Conclusion

In conclusion, our study confirms the interaction between working memory and inhibitory functioning in children and demonstrates these links with planning. Our results are in accordance with those of Grandjean and Collette (2011) with younger and older adults, suggesting an interactive link between working memory and response inhibition by showing that taxing working memory resources increases the difficulty of inhibiting prepotent responses. In our study, adding eggs increased the number of sub-goals but only increased the working memory load if children developed their plan before beginning their motor response, which was not always the case, particularly when the proximity heuristic was activated. As suggested by Kane and Engle (2003), the consistent application of inhibitory processes is constrained by working memory capacities. Our results revealed a developmental effect between 7 and 10 years of age for planning scores and execution time. By contrast, and contrary to our hypothesis, no effect of age was observed concerning the interactions between the number of sub-goals and inhibition of the proximity heuristic. Both 7- and 10-year-old children adapted to changes in the constraints of the task. To study in greater depth the links between inhibition and working memory on the development of planning, it would be interesting to extend the age range and to increase the number of sub-goals without increasing their visual salience.

REFERENCES

Anderson, P. (2002). Assessment and development of executive function (EF) during childhood. *Child Neuropsychology, 8*, 71-82.

Aslan, A., Zellner, M., & Bäuml, K. H. T. (2010). Working memory capacity predicts listwise directed forgetting in adults and children. Memory, 18(4), 442–450.

Conway, A. R. A., Cowan, N., & Bunting, M. F. (2001). The cocktail party phenomenon revisited: The importance of working memory capacity. Psychonomic Bulletin & Review, 8(2), 331–335

Diamond, A. (2002). Normal development of prefrontal cortex from birth to young adulthood: Cognitive functions, anatomy, and biochemistry. In D. T. Stuss & R. T. Knight (Eds.), *Principles of frontal lobe function* (pp. 466–503). New York: Oxford University Press.

De Fockert, J. W., Rees, G., Frith, C. D., & Lavie, N. (2001). The role of working memory in visual selective attention. Science, 291, 1803–1806.

Grandjean, J., Collette, F. (2011). Influence of response prepotency strength, general working memory resources, and specific working memory load on the ability to inhibit predominant responses: A comparison of young and elderly participants. *Brain and Cognition, 77* (2), 237-24.

Hasher, L. & Zacks, R. (1988). Working memory, comprehension, and aging: A review and a new view. In G.G. Bower (Ed.), *The Psychology of Learning and Motivation*. San Diego, CA: Academic Press.

Hasher, L., Zacks, R. T., & May, C. P. (1999). Inhibitory control, circadian arousal, and age. In D. Gopher & A. Koriat (Eds.), Attention and performance XVII, Cognitive regulation of performance. Interaction of theory and application (pp. 653–675). Cambridge, MA: The MIT Press.

Hester, R., & Garavan, H. (2005). Working memory and executive function: The influence of content and load on the control of attention. *Memory & Cognition, 33*(2), 221–233.

Houghton, G., & Tipper, S. P. (1994). A model of inhibitory mechanisms in selective attention. In D. Dagenbach & T. H. Carr (Eds.), Inhibitory processes in attention, memory, and language (pp. 53–112). San Diego: Academic Press.

Huizinga, M., Dolan, C. V., Van der Molen, M. W. (2006). Age-related change in executive function: Developmental trends and a latent variable analysis. *Neuropsychologia, 44*, 2017-2036.

Kane, M. J., & Engle, R. W. (2003). Working-memory capacity and the control of attention: The contributions of goal neglect, response competition, and task set to Stroop performance. *Journal of Experimental Psychology: General, 132*(1), 47–70.

Kane, M. J., Bleckley, M. K., Conway, A. R. A., & Engle, R. W. (2001). A controlled attention view of working-memory capacity. *Journal of Experimental Psychology: General, 130*(2), 169–183.

Lavie, N., Hirst, A., de Fockert, J. W., & Viding, E. (2004). Load theory of selective attention and cognitive control. *Journal of Experimental Psychology: General, 133*(3), 339–354.

Miller, E. K., & Cohen, J. D. (2001). An integrative theory of prefrontal cortex function. *Annual Review of Neuroscience*, 24, 167–202.

Miyake, A., Friedman, N. P., Emerson, M. J., Witzki, A. H., Howerter, A., & Wager, T. D. (2000). The unity and diversity of executive functions and their contributions to complex "frontal lobe" tasks: A latent variable analysis. *Cognitive Psychology, 41*(1), 49–100.

Pennequin, V., Sorel, O., Fontaine, R. (2010). Motor planning between 4 and 7 years of age: changes linked to executive function. *Brain and Cognition, 74,* 2, 107–111.

Redick, T. S., Calvo, A., Gay, C. E., & Engle, R. W. (2011). Working memory capacity and Go/No-Go task performance. Selective effects of updating, maintenance, and inhibition. *Journal of Experimental Psychology: Learning, Memory and Cognition, 37*(2), 308–324.

Roberts, R. J., Hager, L. D., & Heron, C. (1994). Prefrontal cognitive processes: Working memory and inhibition in the antisaccade task. *Journal of Experimental Psychology: General, 123*(4), 374–393.

Wellman, H. M., Sophian, C., Fabricius, W. (1985). The early development of planning: solving comprehensive search problems. In H. M. Wellman (Ed.), *Children's searching: The development of search skills and spatial representation.* Hillsdale, N. J.: Erlbaum.

In: Working Memory
Editor: Helen St. Clair-Thompson

ISBN: 978-1-62618-927-0
© 2013 Nova Science Publishers, Inc.

Chapter 4

CONTRIBUTION OF WORKING MEMORY TO IMPLICIT MOTOR LEARNING IN CHILDREN: A PRELIMINARY REPORT

Alison Colbert and Jin Bo
Department of Psychology, Eastern Michigan University, Ypsilanti, MI, US

ABSTRACT

The ability to integrate individual movements into complex actions is an important developmental achievement. Recent research has highlighted the possible cognitive contributions to motor skill learning. It has been reported that working memory (WM) capacity plays a significant role in implicit sequence learning in young and older adults (e.g., Bo, Jennett, & Seidler, 2011; 2012). Unfortunately however, the contribution of WM to implicit sequence learning in early childhood is still unclear. Thus, the current chapter characterized the development of implicit sequence learning and working memory, and correlated visuospatial and verbal working memory to implicit sequence learning in children. Fourteen neurotypical children ages 6 to 12 years performed a serial reaction time task (SRT) and computerized verbal and visuo-spatial change detection working memory tasks. Preliminary results showed a trend for positive learning ($p = .07$) for this group of children, but age-related changes in sequence learning magnitudes were not found ($p = .38$). Consistent with previous findings in young and older adults (Bo et al., 2012), there was a strong correlation between verbal and visuospatial WM measures ($p < .001$) in children. Interestingly, however, no correlation was found between implicit motor sequence learning and WM measures ($p > .05$), In addition, no age effect for working memory measures was noted ($p > .05$). The current findings seem to support the "development invariant" hypothesis for implicit learning in early development (e.g. Meulemans, van der Linden, & Perruchet, 1998; Reber, 1993). However, additional samples and supplementary measures are needed to better understand the results. Understanding the relationship of working memory to motor sequence learning is critical to conceptualizing the performance of neurotypical children, as well as to provide a foundation for understanding these relationships in children with disabilities in which motor deficits frequently co-occur, such as Attention Deficit Hyperactivity Disorder (ADHD) or Autism Spectrum Disorders (ASD).

INTRODUCTION

Implicit sequence learning refers to the capacity to merge single movements into one smooth, coherent action sequence, while being unable to verbalize the details of the sequence or conscious awareness that learning occurred (Brown et al., 2010; Reber, 1993). A widely used paradigm to investigate implicit sequence learning is the serial reaction time (SRT) task, where participants make key-press responses based on the stimuli appearing successively in a repeating sequential fashion (Nissen & Bullemer, 1987). Typically, implicit learning is concluded when participants demonstrate faster and more accurate responses without explicitly describing what they have learned.

It has been theorized by the "developmental invariance hypothesis" that implicit learning, which appears to recruit more primitive brain regions such as basal ganglia, should demonstrate early functional development and be relatively robust in the face of neurological insults (Reber 1993). It has been demonstrated that implicit learning mechanisms are available very early in infancy (Canfield et al., 1997; Clohessy et al., 2001; Haith & McCarty, 1990; Saffran et al., 1996). Studies on implicit sequence learning in early childhood have shown either no evidence of developmental effects (Meulemans et al., 1998) or very subtle age differences (Thomas & Nelson, 2001). However, other studies using similar sequence learning paradigms (Maybery et al., 1995; Thomas et al., 2004) have clearly shown age-related changes with older children (5-7 years-old) demonstrating stronger learning effects than younger children (10-12 years-old). The contradiction apparent in the existing literature leads us to question what are the factors that may affect the development of implicit sequence learning in early childhood.

Recent studies have suggested that implicit sequence learning, although considered to be a subconscious process, engages cognitive processes such as working memory (WM). The working memory system is conceptualized as a limited capacity, cognitive system that is comprised of the central executive control system (CE), and verbal (VWM) and visuospatial systems (VSWM; Baddeley, 2003). Seidler, Bo, and Anguera (2012) have recently found evidence for the contribution of VSWM to implicit sequence learning. Earlier, Frensch and Miner (1994) reported VWM performance correlated with the magnitude of implicitly learned sequences in young adults. It is known that many cognitive functions, including WM, undergo significant age-related improvement in early child development (e.g. Gathercole, 1999). Interestingly, the relationship between WM and implicit learning in children has been greatly under-investigated.

Developmentally, it has been suggested that the capacity to learn implicitly is important to the acquisition of a wide range of tasks in social (Lieberman, 2000), communicative (Gomez & Gerkin, 1999) and motor domains (Perruchet & Pacton, 2006). Some research indicates that deficits in implicit learning ability may contribute to atypical cognition in neurodevelopmental disorders such as Autism Spectrum Disorders (ASD; Mostofsky et al., 2000) and Attention Deficit Hyperactivity Disorder (ADHD; Barnes et al., 2010). The current chapter therefore had two main goals: 1) to characterize the development of implicit motor learning and WM capacity; and 2) to investigate the influence of WM processes on implicit motor sequence learning in neurotypical children. Based on previous studies which showed age-related increases in implicit sequence learning (Maybery et al., 1995; Thomas et al., 2004), it was hypothesized that both implicit sequence learning and WM would show significant age-related differences. In addition, it was hypothesized that WM capacity would be significantly correlated with implicit sequence learning in neurotypical children.

METHOD

Participants

Fourteen typically developing children (mean age 9.02 years old, range 6 to 12 years; 12 boys, 2 girls) were recruited from the Ypsilanti-Ann Arbor community. Parents or guardians of the participants provided written informed consent, and children assented prior to participation. Participants were given a $10 gift card for participation. The experimental procedures were approved by the Institutional Review Board of Eastern Michigan University.

Procedure

Parents or guardians of the participants completed several behavioral rating scales, including the *Social Skills Rating System* (Gresham & Elliot, 1990), *Conners 3 ADHD Rating Scale* (Conners, 2008), and *Social Responsiveness Scale* (Constantino & Gruber, 2002). Behavioral rating scales were included in order to screen for symptoms of neurodevelopmental disorders in participants. All children performed computerized visuospatial working memory, verbal working memory, and implicit SRT tasks (E-Prime version 2.0, Psychology Software Tools, Pittsburgh, PA).

Working Memory Tasks

We employed the change detection visuospatial (VSWM) and verbal working memory (VWM) tasks reported in Bo et al. (2012). The tasks were originally modified from Luck and Vogel (1997) and Thomason and colleagues (2009). Both computerized tasks began with presentation of a fixation cross, followed by presentation a sample array for 100ms. The fixation cross was then presented again for 900ms, followed by a 2,000ms presentation of the test array. The test array consisted of either the same stimulus as the sample array, or an array with one element changed. After viewing each sequence of arrays, participants were asked to press the corresponding key to indicate if the test array was the same (s) or different (d) from the sample array. In each task, all arrays were arranged along an invisible concentric circle around a fixation cross. For the VSWM task, the arrays consisted of two to eight (array size) colored circles (radius = 1; randomly selected from red, orange, yellow, green, blue, violet, pink, white, black, and brown). On each trial the test array was either the same as the sample array, or changed by one color. Therefore this task relied on the detection of change in color at different locations. For the VWM task, the arrays consisted of letters (size = 1; randomly selected from Q, R, G, B, H, A, N, F, K, Z, S, and V). The sample arrays were uppercase letters, whereas the test arrays were lowercase, forcing participants to encode the letters. On each trial, the test array was either the same as the sample array, or changed by one letter.

Therefore this task relied on detection of letter change. Working memory capacity was estimated using the formula: K = Size of array * (observed hit rate − false alarm rate; Vogel & Machizawa, 2004). The average K across all array sizes was then computed to represent working memory capacity for each participant.

Implicit Sequence Learning Assessment Task

A serial reaction time task (SRT) was adapted from a task described by Brown and colleagues (2010). Participants were seated in front of a computer monitor, and asked to respond as quickly and accurately as possible to a stimulus that appeared in one of four locations on the monitor.

Responses were made on a keyboard with keys marked to correspond to boxes on the computer monitor. Upon pressing the correct key, the stimulus moved to another location. The location of the stimulus followed a 12-step, second-order conditional sequence (312143241342). There were ten blocks of 96 trials per block. The first two blocks were training blocks. The first block started with 20 random trials with no trills or runs for the purpose of familiarization. The rest of the trials in the first block, and all the trials in block 2, 8 and 10 contained an alternative second-order conditional sequence (132341243142). The response-to-stimulus interval was programmed at zero milliseconds. This, combined with the use of a second-order conditional sequence, was used to minimize the use of explicit strategies.

The learning magnitudes were assessed by comparing reaction time (RT) between trials generated in the alternative and learned sequence blocks (i.e., RT differences = RT in block 8 − RT in block 9).

Two sequence awareness tests were used to gauge explicit awareness: generation and recognition tests (details in Bo et al., 2012). In the generation task, participants could start with any key, and were asked to generate a sequence similar to the learned sequence. For the recognition task, participants were presented with 24 three-trial fragments, to which they responded in the same way as the sequence learning task. Half of the elements were from the learned sequence, and half were from the alternative sequence. Following each fragment, the participant was asked to rate how certain they were that the fragment was part of the original learned sequence on a one (certain it was) to six (certain it was not) point likert scale (Shanks & Johnstone, 1999).

RESULTS

Implicit Sequence Learning

Figure 1 shows the group mean RT for each block of the SRT task. The mean RT improvement between blocks 1 and 2 (alternative sequence trials) reflects general practice effects due to becoming familiar with the task. It is interesting that the RT for block 3 (first learning sequence) was faster that block 2.

This might suggest a constant practice effect although more data is needed to explain the phenomenon. To measure implicit learning, we used the RT difference between alternative sequence block 8 and learned sequence block 9. Though this finding did not reach significance ($p = .07$), it approached significance, indicating a trend for learning. Figure 2 shows a noticeable trend for learning that was associated with age, however this finding did not reach significance ($p = .38$). In addition, the overall mean accuracy was 89% across blocks (range = 74-96%) without clear evidence of speed-accuracy trade off.

Two tests were employed to assess participants' awareness after the SRT task. If a child could correctly produce more than five successive positions of the sequence in the generation task (Willingham et al., 1997), and his/her rating scores between the learned and alternative sequence elements were similar in the recognition task, explicit awareness could be concluded. We found that 4 of 14 children generated more than 5 successive elements in the generation task. However, the rating scores between the learned ($M = 2.54$) and alternative ($M = 2.62$) sequences were not different from each other ($P > .10$). Individual data showed that the rating scores were within 1.45-4.3 and 1.46-4.15 for learned and alternative sequences. No significant differences ($t = -.575$, $P = .576$) between two sequences were found in this group of children.

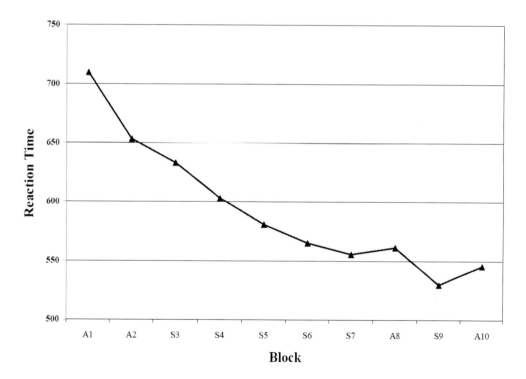

Figure 1. Mean reaction times for each block. Blocks 1, 2, 8 and 9 contained an alternative sequence and blocks 3-7 contained a learning sequence. Note the reaction differences between block 8 and 9.

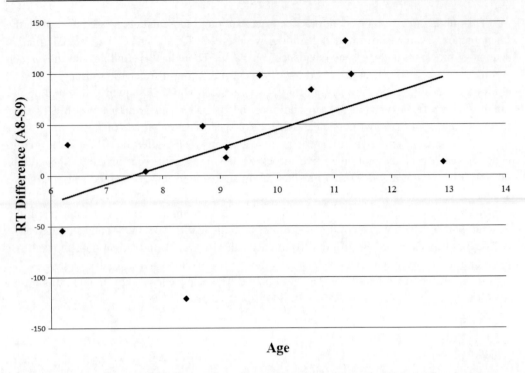

Figure 2. Correlation between age and reaction time difference between block 8 and 9 (i.e., learning outcome for the SRT task).

Working Memory

The mean WM scores were .94 ($SD = 0.67$) for VWM and 1.90 ($SD = .84$) for VSWM. Figure 3 illustrates the relationship between WM capacity and age. A strong correlation between two working memory measures was found ($p < .001$), though no age effect was noted on either WM measure (both $p > .05$). Additionally, it is important to note that data from two participants were excluded in the VSWM analyses due to an erratic response style (i.e., not engaged in the tasks), one participant was unable to complete the VWM task due to inability to recognize letters, and one participant did not complete WM measures.

Correlations between WM and SRT

Figure 4 shows the association between WM capacity and implicit sequence learning. Marginal contributions of both VSWM and VWM on sequence learning (VWM, $p = .10$; VSWM $p = .08$) were found. Correlational analyses between the behavioral assessments and WM also revealed a marginally significant relationship between WM and peer relationships (as measured by the SSRS; both $p < .10$).

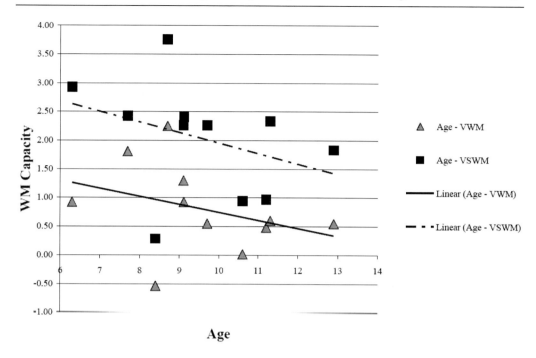

Figure 3. Age effects on VSWM and VWM. Note that neither WM showed significant age-related changes in the current sample.

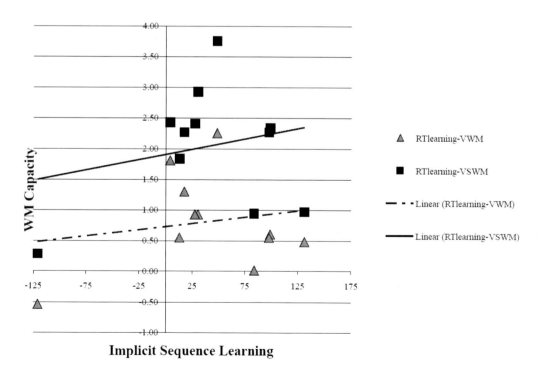

Figure 4. Contribution of WM to the learning outcome for the SRT task. Neither WM measures predicted the reaction time difference between block 8 and 9.

DISCUSSION

We predicted significant age-related improvements on performance in the implicit SRT task and two WM tasks. There was also expected to be a significant correlation between WM and sequence learning measures. The preliminary results, however, did not support either hypothesis. Age-related differences on the sequence learning magnitudes, as well as VSWM and VWM, were not found. Neither VSWM nor VWM was related to the implicit sequence learning in the current sample.

Lack of an age effect on implicit sequence learning was consistent with the developmental invariant hypothesis (Reber, 1993). It has been suggested that implicit learning does not develop steadily over childhood, and instead reaches adult competence early in life. This viewpoint posits early functional development of implicit sequence learning that is set as early as four-years-old (Reber, 1993). Supporting this theory, Meulemans et al. (1998) found no age-related changes in implicit sequence learning ability between children aged six to ten-years-old and adults. Similarly, Thomas and Nelson (2001) found four-year-old children showed substantial implicit sequence learning ability. It is important to note, however, that in the current chapter there was a noticeable trend of stronger learning with increasing age in Figure 2, and potential type II error may exist due to small sample size in this pilot. Thus, the alternative hypothesis suggesting age-related improvement on implicit learning needs to be further explored.

Studies on working memory also debate whether it is a developmental invariant process, or whether it develops throughout childhood. For example, Geier and colleagues (2009) found that although brain processes supporting basic working memory processes are established by childhood, refinement of circuitries also occurs throughout adolescence and into adulthood. Heyes et al. (2012) also found evidence of developmental VWM improvement throughout middle childhood and early adolescence. Thomason and colleagues (2009) highlighted immature VWM and VSWM capacity in children compared to adults, though hemispheric specialization was evident in both. Interestingly, the preliminary analysis on the current data set showed no age-related differences on either VSWM or VWM. Again, the developmental invariant hypothesis of cognitive function seems to be supported.

The significant correlation between VSWM and VWM in the current chapter was consistent with the previous findings in both young and older adults (Bo et al., 2011; 2012). Such results suggest common processes between VSWM and VWM, which support the *domain-general* model (Jarrold & Towse, 2006). In other words, tasks involving spatial storage and maintenance of information cannot be differentiated from those involving verbal or numerical storage (e.g., Oberauer et al., 2000). Recently, it was reported that there were age-related differential declines between VSWM and VWM in aging (Bo et al., 2009; Jenkins et al., 2000), suggesting *domain-specific* processes. Results from the current chapter, however, did not show developmental differentiations between the two WM domains. Here, it is notable that performance on the VSWM measure was overall slightly higher than for the VWM in children. This finding, however, may be solely a function of the task, as the VSWM task may be slightly easier than the VWM task (Bo et al., 2012). Therefore, the question of differential development of VWM and VSWM remains unanswered, and more data is needed.

Although previous studies have shown WM processes are related to implicit learning ability in young adults (Frensch & Miner, 1994: Seidler et al., 2012), other studies have indicated WM processes are important in explicit but not implicit learning (Unsworth & Engle, 2005). The current chapter investigated the influence of VWM and VSWM processes on implicit motor learning ability in neurotypical children. Results revealed only marginal contributions of WM to the magnitude of implicit sequence learning. Previous studies showed that cognitive contribution to learning was especially promising during the early stage (e.g., Bo et al., 2009). Thus, further analysis differentiating stages of learning are planned. Additionally, studies of neurodevelopmental disorders suggest other functions beyond WM may be related to implicit learning (Barnes et al., 2010; Mostofsky et al., 2000; Rapport et al., 2001). Therefore, an exploratory correlation analysis, among multiple cognitive/social measures including inattention, peer relationship, and hyperactivity, was performed. Surprisingly, no relationship was found though correlations between the two WM and peer relationships approached significance.

CONCLUSION

To summarize, the current chapter seems to lend support to the developmental invariance hypothesis of cognitive development, both in the domains of WM and implicit motor learning. In contrast to previous research with adults (e.g. Bo et al., 2009; Bo et al., 2011; 2012), the current data did not support the influence of WM capacity on motor sequence learning in children. However, it is important to note that this is a preliminary report. More analyses and data collection are needed to further understand the current results. A larger sample size would produce stronger power to detect differences. Additionally, the two computerized working memory tasks have not been validated in children. Traditional span tasks will be added to validate the current WM measures.

REFERENCES

Baddeley, A.D., Logie, R., Bressi, S., Della Sala, S., & Spinnler, H. (1986). Dementia and working memory. *Quarterly Journal of Experimental Psychology, 38A,* 603–618.

Baddeley, A. (2003). Working memory: Looking back and looking forward. *Neuroscience, 4,* 829-839.

Barnes, K. A., Howard Jr, J. H., Howard, D. V., Gilotty, L., Kenworthy, L., Gaillard, W. D., & Vaidya, C. J. (2008). Intact implicit learning of spatial context and temporal sequences in childhood autism spectrum disorder. *Neuropsychology, 22(5),* 563.

Barnes, K. A., Howard Jr, J. H., Howard, D. V., Kenealy, L., & Vaidya, C. J. (2010). Two forms of implicit learning in childhood ADHD. *Developmental neuropsychology, 35*(5), 494-505.

Bo, J., Borza, V., & Seidler, R.D. (2009). Age-related declines in visuospatial working memory correlate with deficits in explicit motor sequence learning. *Journal of Neurophysiology, 102,* 2744-2754.

Bo, J., Jennett, S & Seidler, R.D. (2011). Working memory capacity correlates with implicit performance on SRT. *Experimental Brain Research. 214,* 73-81.

Bo, J., Jennett, S & Seidler, R.D. (2012). Age related differences on the correlation between working memory and implicit sequence learning. *Experimental Brain Research, 221*,467-477.

Brown, J., Aczel, B., Jimenez, L, Kaufman, S.B., Grant, K.P. (2010). Intact implicit learning in autism spectrum conditions. *The Quarterly Journal of Experimental Psychology, 63(9)*, 1789-1812.

Canfield, R. L., Smith, E. G., Brezsnyak, M. P., Snow, K. L., Aslin, R. N., Haith, M. M., ... & Adler, S. A. (1997). Information processing through the first year of life: A longitudinal study using the visual expectation paradigm. *Monographs of the Society for Research in Child Development, 62(2)*, 1-145.

Clohessy, A. B., Posner, M. I., & Rothbart, M. K. (2001). Development of the functional visual field. *Acta Psychologica, 106*(1), 51-68.

Conners, C. K. (2008). Conners Comprehensive Behavior Rating Scale Manual. Toronto, Ontario, Canada.

Constantino, J. N., & Gruber, C. P. (2002). The social responsiveness scale.*Los Angeles: Western Psychological Services*.

Destrebecqz, A., & Cleeremans, A. (2001). Can sequence learning be implicit? New evidence with the process dissociation procedure. *Psychonomic Bulletin & Review, 8(2)*, 343-350.

Diperna, J.C., & Volpe, R.J. (2005). Self-report on the Social Skills Rating System: Analysis of reliability and validity for an elementary sample. *Psychology in the Schools, 42(4)*, 345-354.

Frensch, P.A., & Miner, C.S. (1994) Effects of presentation rate and individual differences in short-term memory capacity on an indirect measure of serial learning. *Memory and Cognition, 22*, 9-110.

Gathercole, S. E. (1999). Cognitive approaches to the development of short-term memory. *Trends in cognitive sciences, 3*(11), 410-419.

Geier, C. F., Garver, K., Terwilliger, R., & Luna, B. (2009). Development of working memory maintenance. *Journal of neurophysiology, 101*(1), 84-99.

Gomez, R.L., & Gerken, L. (1999). Artificial grammar learning by one-year-olds leads to specific and abstract knowledge. *Cognition, 70*, 109–135.

Gresham, F.M., & Elliott, S.N. (1990). *Social Skills Rating System manual*. Circle Pines, MN: AGS.

Haith, M. M., & McCarty, M. E. (1990). Stability of visual expectations at 3.0 months of age. *Developmental Psychology; Developmental Psychology, 26*(1), 68

Heyes, S., Zokaei, N., van der Staaij, I., Bays, P. M., & Husain, M. (2012). Development of visual working memory precision in childhood. *Developmental Science, 15(4)*, 528-539.

Jarrold, C., & Towse, J. N. (2006). Individual differences in working memory. *Neuroscience, 139*(1), 39-50.

Jenkins, L., Myerson, J., Joerding, J. A., & Hale, S. (2000). Converging evidence that visuospatial cognition is more age-sensitive than verbal cognition. *Psychology and Aging, 15*(1), 157.

Lieberman, M.D. (2000). Intuition: a social cognitive neuroscience approach. *Psychological Bulletin, 126*, 109–137.

Luck, S. J., & Vogel, E. K. (1997). The capacity of visual working memory for features and conjunctions. *Nature, 390(6657)*, 279-280.

Maybery, M., Taylor, M., & O'Brien-Malone, A. (1995). Implicit learning: Sensitive to age but not IQ. *Australian Journal of Psychology, 47(1),* 8-17.

Meulemans, T., van der Linden, M., Perruchet, P. (1998). Implicit sequence learning in children. *Journal of Experimental Child Psychology, 69(3),* 199-221.

Mostofsky, S. H., Goldberg, M. C., Landa, R. J., & Denckla, M. B. (2000). Evidence for a deficit in procedural learning in children and adolescents with autism: implications for cerebellar contribution. *Journal of the International Neuropsychological Society, 6*(7), 752-759.

Nissen, M. J., & Bullemer, P. (1987). Attentional requirements of learning: Evidence from performance measures. *Cognitive psychology, 19*(1), 1-32.

Oberauer, K., Süß, H. M., Schulze, R., Wilhelm, O., & Wittmann, W. W. (2000). Working memory capacity—facets of a cognitive ability construct. *Personality and Individual Differences, 29(6),* 1017-1045.

Perruchet, P., & Pacton, S. (2006). Implicit learning and statistical learning: One phenomenon, two approaches. *Trends in Cognitive Science, 10,* 233–238.

Rapport, M. D., Chung, K., Shore, G., & Isaacs, P. (2001). A conceptual model of child psychopathology: Implications for understanding attention deficit hyperactivity disorder and treatment efficacy. *Journal of Clinical Child Psychology, 30,* 48-58.

Reber, A.S. (1993). Implicit learning and tacit knowledge: An essay on the cognitive unconscious. Oxford University Press.

Saffran, J. R., Aslin, R. N., & Newport, E. L. (1996). Statistical learning by 8-month-old infants. *Science, 274,* 1926-1928.

Seidler, R. D., Bo, J., & Anguera, J. A. (2012). Neurocognitive Contributions to Motor Skill Learning: The Role of Working Memory. *Journal of Motor Behavior,44*(6), 445-453.

Shanks, D.R., & Johnstone, T. (1999) Evaluating the relationship between explicit and implicit knowledge in a sequential reaction time task. *Journal of Experimental Psychology Learning Memory and Cognition, 25,* 1435–1451.

Thomas, K. M., & Nelson, C. A. (2001). Serial reaction time learning in preschool-and school-age children. *Journal of Experimental Child Psychology, 79*(4), 364-387.

Thomas, K. M., Hunt, R. H., Vizueta, N., Sommer, T., Durston, S., Yang, Y., & Worden, M. S. (2004). Evidence of developmental differences in implicit sequence learning: an fMRI study of children and adults. *Journal of Cognitive Neuroscience, 16*(8), 1339-1351.

Thomason, M. E., Race, E., Burrows, B., Whitfield-Gabrieli, S., Glover, G. H., & Gabrieli, J. D. (2009). Development of spatial and verbal working memory capacity in the human brain. *Journal of cognitive neuroscience, 21*(2), 316-332.

Unsworth, N., & Engle, R. W. (2005). Individual differences in working memory capacity and learning: Evidence from the serial reaction time task. *Memory & Cognition, 33*(2), 213-220.

Vogel, E. K., & Machizawa, M. G. (2004). Neural activity predicts individual differences in visual working memory capacity. *Nature, 428*(6984), 748-751.

Chapter 5

THE DEVELOPMENT OF VISUO-SPATIAL WORKING MEMORY IN CHILDREN

Colin Hamilton
University of Northumbria, Newcastle upon Tyne, UK

ABSTRACT

The aim of this chapter will be to review the empirical research literature which has examined changes in children's employment of visual, spatial mnemonic and attentional processes when carrying out short term visuospatial working memory protocols. The chapter will evaluate the extent to which these processes can be effectively differentiated. An element of this consideration will be to ask whether such a focus directs research and conceptual attention away from working memory as an integrated system, where within a typical short term memory context, the child may bring multiple resources towards successful task performance. The review of the research literature will be embedded within the major theoretical accounts put forward to account for working memory development in children, multiple resources accounts, e.g., Baddeley (1986, 2007), Logie (1995, 2011); the embedded process account, Cowan (1988, 1999) and Oberauer's qualification of this account (Oberauer & Hein, 2011); the continuum model postulated by Cornoldi and colleagues, Cornoldi & Vecchi (2003); and the time-based resource-sharing account of Barrouillet and colleagues (2004, 2009).

OVERVIEW

A common consideration of working memory is that it functions to temporarily maintain and manipulate information in the service of a range of on-going cognitive tasks (e.g. Baddeley, 2012). Within this form of definition, *Visuo Spatial Working Memory* (VSWM) is viewed as the system which maintains and processes non-verbal information. The source of this visuospatial information varies dependent upon the particular conception of working memory, directly from perceptual information, or as activated, interpreted, long term memory content. One particular argument of this chapter will be that this definition of VSWM tends to be driven by the nature of the information media explicit in the working memory task demands, rather than the cognitive resources underpinning task performance. Within a child developmental context, the incongruency between the task surface visuospatial characteristics and the increasing recruitment of a diverse array of cognitive resources emerges as one of the defining characteristics of VSWM task performance change in children.

The chapter content will begin with a brief review of the major theoretical accounts of working memory, conventionally originating within an adult literature, but through the chapter a developmental perspective will be explicitly emphasised. This review will encompass multiple resources accounts (e.g. Baddeley, 1986; 2007, Logie, 1995; 2011); the embedded process account (Cowan, 1988; 1999) and Oberauer's qualification of this account (Oberauer & Hein, 2011); and the continuum model postulated by Cornoldi and colleagues (Cornoldi & Vecchi, 2003).Subsequently, the time-based resource-sharing account of Barrouillet and colleagues (2004, 2009) will be considered. The identification of these theoretical frames will be followed by a summary of the research literature which has aimed to identify changes in the efficacy of visual, spatial and executive or attentional processes within VSWM. The thrust of this summary will be towards conventionally labelled 'simple' span tasks, rather than the more attentionally engaged 'complex' span literature. Relevant evidence from processing speed accounts and developmental research with pictorial stimuli will be employed to demonstrate the potential for complex representation of visuospatial information. The final component of the chapter will attempt to interpret the research literature through an explicit consideration of the theoretical frames and research issues.

SOME THEORETICAL FRAMES

Multiple Resources Accounts

The original multiple resource or component working memory account suggested by Baddeley and Hitch (1974) envisaged three discrete resources, a *Central Executive* control process, a *Phonological Loop* and a *Visuo Spatial Sketch Pad* (VSSP). This latter process was envisaged as "…retaining and manipulating images, and is susceptible to disruption by concurrent spatial processing…" (Baddeley, 1986). In more recent accounts, the VSSP has been placed explicitly within the context of perceptual and long term memory processes (Baddeley, 2000, 2012). In these recent formulations, in addition to the domain specific slave processes there is also an *Episodic Buffer* process which retains integrated material or chunks which may be amodal or multidimensional in nature. Figure 1a identifies the architecture of the VSSP, emphasising the perceptual input to the store (a gateway account) and its interface with the Episodic Buffer. The VSSP is also subject to direct support and scaffolding from Long Term Memory (LTM) visual semantics and indirectly from LTM verbal semantics.

An alternative multiple resource model was proposed by Logie (1995, 2011). In this account there was an articulation of the VSSP component of the Baddeley (1974, 1986) model, a *visual cache* process for the storage of visual material and an *inner scribe* process for dynamic spatial representation and the active rehearsal of the passive visual material (See Figure 1b below) . This model demonstrated a striking symmetry in function between the verbal and non-verbal components of the model. However, one should note that perhaps the most fundamental difference between these two accounts is that in Logie's proposal, all of the information in working memory has gone through interpretation by the LTM (Knowledge Base) *prior* to entry into working memory. This would be conceived of as a *workspace* account rather than a gateway account (Logie, 1996). Thus within this perspective, the visual content of VSWM is likely to be in the form of fully integrated, fully bound objects. Whereas in the Baddeley account (Figure 1a), feature representation and bound representations may co-exist within the VSSP, e.g. colour, shape along with the integrated object (see also Baddeley, Allen & Hitch, 2011, Figure 6, p 1398).

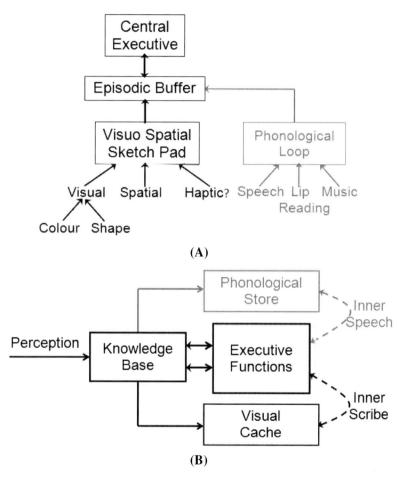

Figure 1. a) The multiple resource account of Baddeley (2012); b) The multiple resource account of Logie (2011).

Embedded Processes Accounts

These accounts tend to be less modular than multiple resource accounts, sometimes labelled single (attentional) resource accounts. The earliest such model is the Cowan (1988, 1999, 2008; Engle, Kane, & Tuholski, 1999) *Embedded Processes* account. In this account working memory representation is seen as activated LTM sensory and categorical information. This activation level is above the typical level of LTM information; however information that is within the *Focus of Attention* is seen to have the highest level of activation (see Figure 2a below). This conceptualisation has been likened to the distinction between secondary and primary memory (James, 1890). The key process at work here is the attentional modulation within the focus of attention, what Cowan, Fristoe, Elliot, Brunner and Saults (2006) have labelled as the *scope* or *capacity* of attention. Unsworth and Engle (2007) have argued for a second attentional resource, associated with retrieval from secondary memory, activated LTM outside of the focus of attention (retrieval from within the focus is assumed not to require active processes). If comparisons can be made between this account and Baddeley's multiple resource account then one such comparison could be equating the focus of attention with a central executive process (as Cowan & Alloway, 2009, actually did, shown in Figure 2b).

Oberauer and Hein (2012) have modified the embedded processes by distinguishing between a broad focus of attention, of 3-4 items, in contrast to a component with a narrow focus of attention, specifically 1 item. This account accommodated prior experimental evidence which suggested that one chunk or item had privileged access and status. This is demonstrated in Figure 2b, which illustrates a schematic network in long term memory, with 5 items currently activate (shaded circles), and 3 items (a, b, & c) under the broad focus of attention, but 1 item receiving privileged access status (a).

Integrative Accounts

The two broad approaches identified above tend to emphasise modality specific and controlled attentional resources respectively. One account, the Continuum Model (Cornoldi & Vecchi, 2003) emphasises both modality specific stores and controlled attention. In this model, shown in Figure 2c, there is a vertical continuum varying in the degree of attentional control needed to carry out the task at hand. At the passive end of the continuum, the lower portion of the figure, where maintenance and storage only is required, the authors envisage quite discrete modality specific processes. As task demands increasingly require more controlled attention, a movement up the continuum, the resources recruited become increasingly amodal and generic. Whilst this appears in many ways similar to the Baddeley multiple resource account discussed above, it makes more explicit the notion of a flexible allocation of attention to task completion (Kahneman, 1973) and consequently undermines the naive categorisation of tasks into *Simple* and *Complex*. The majority of VSWM tasks will inevitably vary in their requirements for maintenance and controlled attention resources.

Whilst the theoretical accounts identified above tend to be formulated and considered within a young adult context, the content of this chapter will also consider accounts specifically associated with lifespan development, e.g. the putative contribution of inhibitory processes (Hasher, Lustig & Zachs, 2007), and the contribution of Speed of Processing (SOP, Fry & Hale, 1996; Kail & Salthouse, 1994). In particular, a more recent account the Time-Based Resource-Sharing model proposed by Barrouillet and colleagues (2004, 2011) will be considered. This framework has been successfully applied within a child development context (e.g. Camos & Barrouillet, 2011).

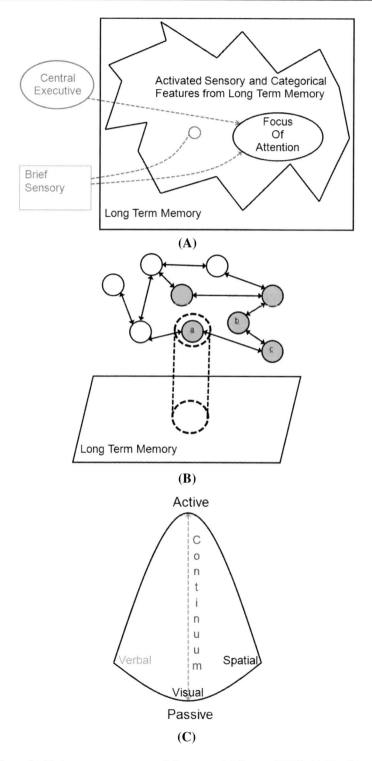

Figure 2. a) The embedded processes account of Cowan and Alloway (2009); b) The three embedded component model (Oberauer & Kliegel, 2012); c) The Continuum Model (Cornoldi & Vecchi, 2003).

THE DEVELOPMENT OF VISUAL AND SPATIAL WORKING MEMORY TASK PERFORMANCE

Visual Working Memory Research Findings

A large number of research protocols have been employed in order to assess visual working memory in children, the stimuli have ranged from visual matrix patterns (Della Sala, Gray, Baddeley, Allamano & Wilson, 1999; Phillips & Baddeley, 1971; Pieroni, Rossi-Arnaud, & Baddeley, 2012), coloured patterns arrays (Luck & Vogel, 1997; Riggs, McTaggart, Simpson, & Freeman, 2006), to static and dynamic arrays (Mammarella, Pazzaglia, & Cornoldi, 2006; Pickering, Gathercole, Hall, & Lloyd, 2001). This current section will focus on a number of research protocols, and then in the final section discuss the implications of the results from the wider empirical context. It is worthwhile re-iterating that the vast majority of these protocols were designed for young adult research rather than as a tool for understanding individual differences associated with the lifespan and child development in particular.

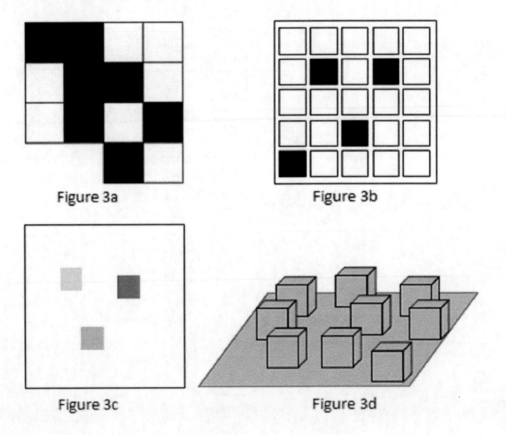

Figure 3. Visuo Spatial Working Memory Task Stimuli.

Three of the common visual working memory protocols are illustrated in Figure 3 above. The visual matrix or Visual Pattern Test (VPT) protocol is shown in Figure 3a, typically a matrix pattern where 50% of the cells are black, 50% are white. The pattern is shown for a period of time (encoding duration) then retained (maintenance interval) then retrieved either though a recall context or recognition process. There is variability in the protocol formats and proportion of black cells may vary, as in the matrix pattern shown in Figure 3b, a protocol employed by Pieroni and colleagues (e.g. Pieroni et al., 2011; Rossi-Arnaud, Pieroni & Baddeley, 2006). In addition, encoding time and maintenance may vary between studies, or actually be the focus of experimental manipulation (e.g. Wilson, Scott, & Power, 1987). A common element is the use of span measurement where the number of black cells is increased until the participant failed to meet the criterion. It should be noted that there is also variability in how one measures the span level, with implications for the reliability of the data (Friedman & Miyake, 2004; St Clair-Thompson, 2010). Another commonly employed protocol in adult research, the visual array procedure of Luck and Vogel (1997) has also been employed within a child developmental context. The conventional array format is shown in Figure 3c, the array size can contain between 1-6 items, is typically briefly presented (< 1 second) and the maintenance interval also tends to be brief (0.5-3seconds). This protocol has been employed in a number of child developmental studies, e.g. by Cowan et al. (2005) and Riggs and colleagues (Riggs et al., 2007; Riggs, Simpson & Potts, 2011). Visual working memory capacity in this protocol was measured in terms of k, derived from array size, and the hits and false alarm rates in task performance (see Pashler, 1988).

Figure 4 identifies the general developmental trajectory for visual matrix span development in children. The VPT span data points are drawn from three studies, Wilson et al. (1987), Logie and Pearson, (1997) and Miles, Morgan, Milne and Morris (1996). Despite differences in the precise characteristics of their respective protocols the combined data neatly fits a developmental function which indicates a rapid increase in VPT span performance from the age of 5 years through to 15 years of age. The graph also reveals another interesting characteristic of the research findings, that the 15 year old children are performing at level equivalent to adults in two of the studies (Wilson et al. and Miles et al.). However, this performance level is well above the typical young adult level in the VPT noted by Della Sala, Gray, Baddeley and Wilson (1997). In their sample of 345 adults the mean performance was 9.08. This discrepancy could be due to the recall protocol in the Della Sala et al.task as opposed to the recognition data in Figure 4 (see Logie & Pearson, 1997, for a direct comparison of span performance in the two retrieval contexts). Other differences in terms of differences in progression criteria, encoding or maintenance duration, or of pattern configuration in the form of complexity could also account for this difference and the latter will be discussed at length below.

Also shown in Figure 4 is the data from the Riggs et al. (2007) study which employed the visual array format adopted from Luck and Vogel (1997). Three age groups were employed in this study, 5, 7, and 10 years of age. The developmental function overlaps with the VPT function only when the two scales are equated for adult performance level. Typically young adults achieve a k score of ~ 4 items, a quantitatively similar level of performance to that with the matrix employed by Pieroni et al. (2011). However, the k value is significantly smaller than the span level typically found in the visual matrix procedure. Identifying the factors which could account for this discrepancy lies at the heart of the discussion in the next section.

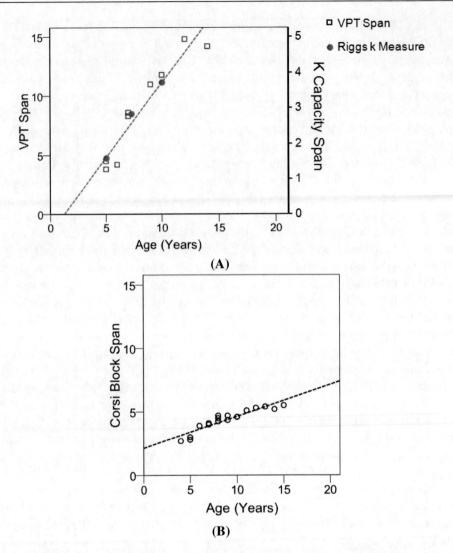

Figure 4. a) The Developmental trajectory associated with the VPT Span and k Capacity measures; b) The developmental trajectory associated with the Corsi Block task.

The developmental work by Pickering et al. (2001) explored a particular characteristic of the array during the encoding of visual matrix patterns, whether the black cells were presented simultaneously on the screen (static version) versus the appearance of individual cells in a sequence of screen shots (dynamic version). The age associated development, as measured by number of trials correct, demonstrated a large increase between the ages of 5-10 years, congruent with the improvement shown in Figure 4 with VPT span data. However, the improvement in the dynamic version, although significant, was not as large as that of the static version. This pattern of *developmental fractionation* (Baddeley & Hitch, 2000) was mirrored in static and dynamic versions of a maze task, and led the authors to conclude that static and dynamic sub component processes were present in visual working memory. Their conclusion was congruent with the spatial-simultaneous versus spatial-sequential differentiation in the Continuum model (Cornoldi & Vecchi, 2003). However as Pickering et al.conceded, their dynamic matrices task possessed the characteristics of the Corsi Blocks task (see Figure 3 d), a task conventionally taken to be a measure of Spatial working memory (Della Sala et al., 1999; Logie, 2011). The next section will consider the research findings associated with Corsi and other measures of spatial working memory.

Spatial Working Memory Research Findings

One of the major prototypical spatial working memory tasks conventionally employed in developmental research has been the Corsi Blocks task (Milner, 1972). The Corsi task (see Figure 3d above) was originally constructed as a complementary task to the digit span task within a clinical context, in order to reveal hemispheric specialization; more specifically to make demands upon the temporal lobe and hippocampus in the right hemisphere (Milner, 1972). It shares two major elements in common with the digit span task: the sequential presentation of the item (in the Corsi task, the sequential tapping of individual blocks) and the demand for accurate serial order in the recall of phase of the task. More recently, researchers have employed PC generated Corsi or spatial span procedures, but as with the VPT task, variability in the protocol exists across studies (Berch, Krikorian & Huha, 1998). The most common outcome measure is the span measure, typically ascending, whereby the number of taps is increased until the participant fails to meet the progression criterion.

The data from 3 studies are superimposed in Figure 4b, data from Orsini, Capitani, Laiacona, Papagno, and Vallar (1987), Isaacs and Vargha-Khadem (1989), and Logie and Pearson (1997). The data points again fit onto the regression function despite differences in protocols. The Corsi Block Span scale in Figure 4b has been adjusted to match that in 4a, where the VPT is plotted. This allows a more direct comparison and suggests that the age associated improvement in the Corsi is not as large as that of the VPT developmental change. This was addressed directly by Logie and Pearson who revealed a significant interaction in the task (VPT vs Corsi) by age group analysis. This was taken by the authors as an indication of *developmental fractionation* (Baddeley & Hitch, 2000) and evidence for differentially age associated development in two processes; namely the visual cache and inner scribe (Logie, 1995, 2011). Interestingly, this pattern of results is similar to that reported by Pickering et al. (2001), however, the demonstration of developmental fractionation was interpreted differently (see above). One should also note that there is evidence of concurrent validity for the Corsi using other measures of spatial memory in children (e.g. Pentland, Anderson, Dye & Wood, 2003)

Postma and De Haan (1996) proposed a three-process account of object location or spatial memory; a *position only* process which allowed the precise spatial location of an object, based upon a coordinate spatial relations process; an *object-to-position* process which allocates object to locations, a process which would draw verbal as well as visuospatial working memory processes; and a *combined* process which integrates the two processes above. Cestari, Lucidi, Pieroni and Rossi-Arnaud (2006) investigated the age associated trajectories in 6, 8 and 10 year old children. They found that position only and object-to-location processes showed early improvement, between 6-8 years, however, the combined process developed later between 8-10 years. Thus, their finding suggested the presence of significant age associated improvements across the age groups, but with interesting differences in trajectories of these processes.

ISSUES IN THE INTERPRETATION OF VISUAL AND SPATIAL WORKING MEMORY RESEARCH FINDINGS

A fundamental question of any of these studies is the extent to which the research question and data allow one to identify which processes are actually changing in child development (Cowan & Alloway, 2009; Hamilton, Coates & Heffernan, 2003). If one employs the frameworks in Figure 1 as a filter for this interpretation, complexities are immediately identified in the process. As Conway et al. (2006) suggest, the first issue concerns the diversity of conceptual framing employed in the working memory literature and how this undermines the mapping '…from constructs to mechanisms to measures…' (p4). An example of this is evident in the labelling of VSWM tasks (as well as verbal tasks) as 'simple' and 'complex'. Within a multiple resource account such as Baddeley (1986) and Logie (1995) simple tasks would presumably make exclusive demands upon the 'slave' systems such as the VSSP, Visual Cache or the Inner Scribe. Within an embedded processes account (e.g. Cowan, 1999) or continuum account (Cornoldi & Vecchi, 2003), simple tasks would have the capability of demanding executive or attentional resources, but less so than those tasks labelled as complex. Thus, tasks such as the VPT or Corsi may recruit visual and verbal semantics (Baddeley, 2000) or processes such as recoding, chunking, rehearsal (Engle et al., 1999). There is extensive dual task evidence in adults to indicate that tasks such as the Corsi and VPT are impaired when secondary task interference targets executive or attentional resources (Fisk & Sharp, 2003; Rudkin et al., 2007; Vandierendonck et al., 2004; Vergauwe et al., 2009). More importantly research has indicated that visuospatial task performance in children can be affected by attentional and non-domain specific interference. In tasks designed to mimic the cognitive demands of the VPT and Corsi tasks, Hamilton et al. (2003) employed verbal fluency (LTM retrieval) interference during the maintenance phase and observed that both visual and spatial task performance was significantly impaired. The impact of this verbal secondary task suggests a role for amodal resources in these VSWM task demands. Thus, even these simple tasks appear to require maintenance *and* processing resources, as indicated in early considerations of the VSSP (e.g. Baddeley, 1986).

The Contribution of Processing Speed to Task Performance

Early accounts of individual differences associated with child development (and adult ageing) have emphasised the importance of processing speed as a key source of change. Case, Kurland and Goldberg (1982) suggested that total capacity (*M capacity*) remained constant over development, however they suggested that processing speed may improve leaving more space for storage. This may account for why Hamilton et al.(2003) also noted that a simple measure of processing speed, articulation rate, was significantly correlated with the younger children's VSWM task performance in their study. The processing speed account was qualified by Towse, Hitch and Hutton (1998) in their *task-switching* account of working memory development in children. They accepted that processing speed could impact upon span performance but argued that this was because information being maintained in working memory was subject to rapid decay and therefore quick processing meant that children could access this decaying information more promptly, and were thus more efficient in reconstructing the information. Hitch, Towse and Hutton (2001) went on to identify that simple accounts of processing speed could not account equivalently for discrete complex span procedures, e.g. operation span versus counting span (see also Cowan, Elliot, et al., 2006). More recently Barrouillet and colleagues (e.g., Barrouillet & Camos, 2011; Portrat, Camos & Barrouillet, 2008) have suggested that both resource sharing and trace decay contribute to working memory task performance, they labelled this a *Time-Based-Resource-Sharing* (TBRS) account (Barrouillet, Bernardin & Camos, 2004). They argued that attention was required for two key processes in working memory, to carry out the processing demands and to actively maintain (or reactivate) the information being maintained.

The Barrouillet argument differed from Towse et al. by implying that attention could be allocated to refreshment or reactivation ant any point during the processing phase of a task if the cognitive load of the processing component did not demand sustained attention. Thus, they demonstrated that it is not time of the processing phase per se which matters, but the extent to which attention is occupied during this phase. In a recent child developmental study Gaillard, Barrouillet, Jarrold and Camos (2001) provided evidence that in order to make working memory span performance in 8 and 11 year olds equivalent, not only had processing efficiency to be controlled for but also the time available for reactivating memory traces needed to be equated. This study identifies the importance of the active maintenance of information in working memory and indicates that in younger children this is attentionally demanding and requires more time. The focus upon refreshment and rehearsal within children is developed below.

Visual and Verbal Representation in VSWM Pictorial Task Performance

Recent research has focused upon the distinction between rehearsal processes and refreshment processes in both adults (Camos, Mora & Oberauer, 2011) and in children (Camos & Barrouillet, 2011; Tam, Jarrold, Baddeley & Sabatos-Devito, 2010). The work of Camos and Barrouillet (2011) looked at changes in children aged 6 and 7 years in how they maintained verbal labels related to pictorial stimuli. Their procedure manipulated both cognitive load and the processing duration, and is demonstrated in Figure 5 below.

The standard procedure involved being exposed to animal pictures and then subsequently recalling the names of the animals, thus the verbal recall protocol was explicit in the procedure. In the baseline condition, shown in 5a, the children were exposed to an animal stimulus (bear) for 2000ms, then had to name the colour of a smiley face, within a 4000ms processing period, before being exposed to the second animal stimulus (cat). In the heavy cognitive load condition, Figure 5b, between the presentation of animal stimuli, the children had to name the colour of two smiley faces within a 4000ms processing period, thus the cognitive load was doubled in the condition, but the processing duration was the same as in the baseline condition. In the final condition, the long processing duration condition shown in Figure 5c, the children, between the presentation of animal stimuli, had to name the colour of two smiley faces within a 8000ms processing period. This condition therefore had the same cognitive demand as the baseline, but a processing duration twice as long as the baseline.

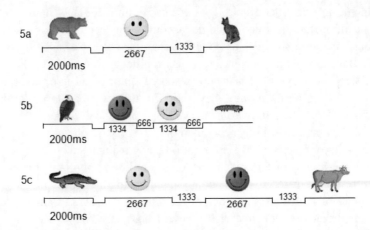

Figure 5. Investigating the impact of cognitive load on attentional refreshment, derived from the Camos and Barrouillet (2011) protocol.

The Camos and Barrouillet (2011) findings suggested that for the younger children aged 6yrs, the change in cognitive load did not affect performance, but the increase in processing duration did impair performance. In contrast, with the older children aged 7yrs, only the cognitive load manipulation affected task performance. The authors argued that this pattern of results reflected a qualitative shift from a passive rehearsal process in the younger children to an active maintenance or refreshment process in the older children, when cognitive load manipulation led to impaired performance. They also suggested that the results in the younger children could be accounted by the task-switching account of Towse, Hitch and Hutton (1998), whereas the cognitive load effect observed in the older children could be more readily explained with reference to their Time-Based-Resource-Sharing account. The authors also interpreted their results in terms a process of temporal decay rather than interference.

Much of the research identified above regarding processing speed involved complex span procedures, and much of the research on refreshment and rehearsal has employed verbal material, and thus its relevance to Visuo Spatial Working Memory task performance may not at first glance appear obvious. However, there has been a long lasting debate within the VSWM developmental literature concerning the nature of visual and verbal representation with visuospatial material such as objects. The early work of Hitch and colleagues looked in detail at the presence of multiple representations with pictorial stimuli (e.g. Hitch, Halliday, Schaafstall & Schraagen, 1988; Hitch, Woodin, & Baker, 1989). Hitch et al. (1988) contrasted visual working memory in 5 and 10 years olds for line drawings of common objects. Through a systematic set of experimental procedures investigating serial position curve attributes, visual similarity and retroactive inhibition effects they established that with pictorial stimuli, the 10 year olds were more reliant on phonological representation. However, even in the youngest children, phonological representation could be employed, and in the older children, some visual representation was still employed (Hitch et al., 1989).

The complexity of pictorial representation through childhood is illustrated in the work of Palmer (2000) who manipulated visual and phonological similarity in tasks demanding the maintenance of line drawing pictorial stimuli similar to the Hitch et al. (1988, 1989) procedures. Palmer looked at the representation of the pictorial images in children aged from

3 to 8 years. She observed that the youngest children were not, in the main, susceptible to either visual or phonological similarity effects. However, from the age of 3 to 6 years, a visual coding strategy, evidenced by a visual similarity effect (VSE), was present. In the 7 year olds, both visual and phonological similarity effects (PSE) were apparent, some children demonstrated only a VSE, some only a PSE, and some both. A second study found a similar pattern of complexity across a 3 year sequential period. Palmer concluded a particular sequence of coding occurred, an early phase where no strategic process is at work, performance being facilitated by LTM activation associated with the pictorial stimuli, followed by visual then phonological coding, where a central executive process allocates attention to temporarily activated LTM representations. Palmer argued that in older children, executive processes would inhibit the visual representation leading to phonological dominance. The emphasis in this study was of VSWM operation within a workspace account.

Henry, Messer, Luger-Klein and Crane (2012) identified a major issue with the visual/phonological research identified immediately above, the implicit and explicit cueing of phonological representations of the pictorial stimuli. Identifying research which demonstrated PSE in very young children, Henry et al. discussed how the observation of children's spontaneous use of phonological recoding may be undermined by the child's explicit verbal labelling of the stimuli, prior to the onset of the procedure, and the requirement for spoken recall at retrieval. Their procedure began with the experimenter naming the objects, and then subsequently employed a sequential presentation of a set of pictures, followed at retrieval by a non-verbal response, pointing in the same sequence to the target stimuli on a response board. The results suggested that the 4-5-year olds did not show any evidence of visual, phonological, or semantic similarity effects, nor word length effect. However, the 6-8-year olds did demonstrate PSE and word length effects. A second experimental replication, using potentially a more sensitive measure, the total number of items correct being the dependent variable (Friedman & Miyake, 2004; St Clair-Thompson, 2010) found a slightly different pattern with a significant VSE in the high achieving sub-group (cluster) of the sample. A third experiment confirmed the presence of PSE and word length effects in children with high task performance. Henry et al. (2012) concluded there was strong evidence for phonological coding of the pictorial stimuli, evidence by PSE and word length effects; a recoding process which develops gradually in children aged between 5-6 years. In 7-year olds, performance was related to the children's self-report of phonological coding. There was only weak evidence for visual representation, and marginal evidence for dual visual/verbal coding.

What is consistent in the developmental research employing pictorial stimuli is the finding of increasing phonological representation of the stimuli through childhood, with the declining importance of visual representation. However, it clear that issues associated with specific protocols may constrain one's interpretation of the findings. Henry et al. (2012) point to the potential confounds arising from a procedure where the child explicitly identifies the names of the pictorial objects and has to provide a spoken response at retrieval. However, in a context where the familiar objects are presented, it is clear that there is always the potential for visual and verbal LTM semantics to be available to the child, the question then becomes at which stage will the child be able to spontaneously make us of these LTM resources? Thus, any naming of the pictures, by participant or experimenter may merely emphasis the verbal semantics associated with the stimuli, regardless of the retrieval protocol. An additional issue associated with the procedures employed in the pictorial research has been the sequential presentation of the stimuli.

Whilst in some research, simultaneous versus sequential stimulus presentation appears to have no major impact upon span (Cowan, AuBuchon, Gilchrist, Ricker & Saults, 2011), the nature of the task demand does change significantly, particularly if the recall requires correct serial order (Avons, Ward & Melling, 2004; Jones, Farrand, Stuart & Morris, 1995). There is perhaps the requirement for more controlled attention as the representation is updated after each event. Extensive research has suggested that with sequential presentation of visuospatial material the typical recency effect is restricted to the last item (e.g. Avons et al., 2004; Phillips & Christie, 1977). An argument made by Phillips and Christie is that recency item and pre-recency items reflect short-term and long-term memory representations respectively. Interestingly, in the context of child studies, Phillips and Christie noted that the single item recency occurs with both short and long lists, the shorter length necessarily being employed in child studies. The presence of this recency component in sequentially presented protocols is evident in the Hitch et al. (1988) study. Whilst in adults, the retrieval context, backward retrieval or probed stimulus does not necessarily influence task performance (Phillips & Christie, 1977); the retrieval protocol with young children may be more problematic. Hitch et al. (1988, study 2) demonstrated that forward or backward recall of sequentially presented stimuli substantially impacted upon the serial position curve (SPC), the recency effect almost completely disappeared with forward recall in both the 5 and 10-year old groups. This would suggest that the short term component of visual memory is minimised in the forward recall protocol. In addition Hitch et al. noted (study 3) that visual retroactive interference had its major effect predominantly upon the recency item. Thus, with children, investigation of a visual representation in short term memory may be seriously compromised by forward recall within a sequential presentation protocol.

The initial developmental research literature discussed above focused upon tasks such as visual matrices, Corsi blocks and visual arrays where the explicit verbal characteristics of the stimuli are minimised, thus, it could be possible that in these tasks, verbal recoding is minimised. However, even with the VPT stimuli, verbal semantics can be employed to give meaning to the black and white cell configurations (Brown, Forbes & McConnell, 2006). It is clear from a range of paradigms that participants will attempt to reduce the complexity of the stimulus by organising and chunking the stimuli in order to give more meaning in the form of verbal and visual semantics to the pattern. Sun, Zimmer and Fu (2011) identified two elements associated with complexity in visual patterns; *physical complexity* associated with low level or perceptual cues explicit in the pattern and *perceived complexity* associated with the relevant expertise brought to the study by the participant, this would include the notion of visual and verbal semantics identified above. Research has established the importance of low level cues in the form of symmetry, grouping, and configuration upon VSWM task performance (e.g. De Lillo, 2004; Kemps, 2001; Rossi-Arnaud, Pieroni, Spataro & Baddeley, 2012). Whilst it is possible that these low level cues could be utilized effectively early in child development, the strategic utilization of verbal and visual LTM semantics is likely to demonstrate progress through childhood.

Although it has been suggested in adults that verbal interference does not necessarily impair VSWM task performance (Morey & Cowan, 2004), Palmer (2000) has argued that in children, the use and integration of visual and verbal information in VSWM task performance was likely to arise though attentional processes. The integration of features within or between modalities may be considered to be a primary or primitive role of an executive process 'dual task' function (e.g. Baddeley, 1996), however there is evidence to suggest that the efficacy of some elements of binding increase through childhood (Brockmole & Logie, 2013; Cestari et al., 2006).

Issues in Identifying VSWM Capacity Changes in Childhood

The content immediately above has highlighted some of the putative issues when trying to understand what changes in VSWM task performance in children. This section will focus upon research which has attempted to disentangle the potential factors underlying VSWM task performance in children.

A starting point is to consider how and where attentional processes could contribute to VSWM task performance. If working memory is seen as a workspace, where content is interpreted prior to entry into VSWM, then attentional processes could potentially constrain the entry process into VSWM; an *encoding* impact. Vogel, McCollough and Machizawa (2005) provided evidence in adults that a key process in working memory efficacy is the inhibition of (goal) irrelevant stimuli, so preventing them from loading working memory. Cowan et al. (2006) have labelled this a *control of attention* process. The focus of attention (see Figure 2a) would act to maintain a number of activated items or chunks; Cowan et al. labelled this aspect the *scope of attention*, and is at work during the *maintenance* phase of working memory. However, other attentional processes could also be at work during the retention phase, organising the information, e.g. forming complex chunks informed by both visual and verbal semantics, and these could be considered further attentional control processes. The presence of such control processes may lead to an overestimate of the scope of attention (Cowan, 2011) In addition, there is also potential for attentional influences at the point of *retrieval*, both in terms of the retrieval processes per se (Unsworth & Engle, 2007) and the retrieval context (e.g. single probe versus full screen recognition, Wheeler & Treisman, 2002). It could be argued that the impact of the retrieval context is less considered in the research literature than the other two phases of the working memory task process.

Recent work by Cowan et al. (2011) has attempted to disentangle the contribution of encoding and capacity in VSWM development. These authors established a procedure whereby children had to allocate attention to a subset of objects (e.g. circles over triangles) within an array. This manipulation allowed the authors to consider what proportion of target stimuli were maintained and the number of non-targeted stimuli remembered, the latter an indication of inefficient attentional filtering at encoding. The study employed three age groups, 6-9 years, 11-13 years, and 18-21 years. The results suggested that the relative attentional filtering efficiency of the youngest age group was equivalent to that of the adult group. However, despite the lack of age difference in this attentional control process, there was an age associated improvement in the task performance. The authors interpreted this as indicating an improvement in the visual memory capacity of the children, i.e. an increase in the scope of attention. It is difficult to consider where in the Visual Patterns Test, Corsi task, or in the original Luck & Vogel (1997) visual array protocol, an inhibitory filtering process would be as important for task performance, as all of the information presented is meant to be remembered and retrieved. However, there may be a temporal constraint within the encoding process associated with attentional disengagement (Fukuda & Vogel, 2011) and should this occur in a child context with brief presentation times then performance in any VSWM task will suffer.

An important characteristic employed in the Cowan et al. (2011) study was the use of verbalization in some of the conditions, either the target object's colour was identified, or the child had to say 'wait' after each item was presented. The former condition, naming the colour, could be seen as a way of implicitly cueing the participant to consider the use of verbal labelling, and indeed in both child groups led to an improved k-capacity score. However the 'wait' utterances manipulation was viewed by the authors as a protocol mimicking articulatory suppression, and indeed in *all* 3 age groups this reduced the span level. These results suggest that even with nonverbal stimuli, verbal semantics can be employed by children, but more importantly for the authors, that in the absence of these phonological labels, capacity continued to increase through childhood. It is clear that in the VPT, Corsi and Luck and Vogel protocols, the presence of verbal semantics could inflate performance.

Perhaps more important for these conventional VSWM protocols is the extent to which visual, verbal, and configurational cues could lead to elaborated organization and chunking in order to reduce the visual complexity, a process of compression (e.g. Kemps, 2001; Rossi-Arnaud et al., 2012). This would lead to improved VSWM task performance, not necessarily because of an increase in the scope of attention, the number of chunks, but rather that compression has led to a richer quality, i.e. more information within the chunk. Research by Cowan, Hismjatullina et al. (2010) has attempted to disentangle the contribution of capacity and compression processes to changes in VSWM performance in children. In their experimental protocol, pictorial stimuli were initially associated as pairs or triplets (as opposed to singletons) in a familiarisation phase prior to becoming elements of a memory procedure. The results suggested that the extent to which adults developed rich (compressed) chunk representation was greater. However, whilst controlling for this improvement the authors were able to observe that from the 8 year-old group to adulthood, there was an increase in the number of chunks that were utilized, thus capacity was still improving. This was also the case when compared with adults who were subject to articulatory suppression. Therefore, although visual (and verbal) semantics can improve VSWM task performance through improving the richness or quantity of information within a chunk, the scope of attention demonstrates independent developmental improvement.

This suggests that in the matrix stimuli of Kemps (2001) and Rossi-Arnaud et al. (2012) and the Luck and Vogel (1997) array formats, there is scope for some development of chunk compression efficacy through childhood, potentially inflating an accurate measure of capacity. This is more clearly seen in the performance of children in the conventional VPT task where by the age of 5 years, children are remembering more black cells than the typical adult visual memory capacity level (see Figure 3 above). What is required in all of this developmental research is the development of protocols where chunk compression efficacy can be identified and a distinction made between perceptually driven visual organisation (physical complexity in the pattern) and perceived complexity (Sun et al., 2011), where on-line compression activity may occur in the absence of explicit LTM priming which was the case in the Cowan et al., (2010) study.

Beyond a Quantitative Approach to VSWM Development

The emphasis in much of the conventional VSWM research identified above and in the pictorial stimuli research has been upon the VSWM capacity or scope of attention associated with child development. This quantitative approach has been complemented by research which has looked at how children may recruit any cognitive resource pertinent to task completion. This research has made use of the functional architecture of working memory, whereby processes endogenous to, or exogenous to working memory may be recruited (Baddeley, 2000; Hamilton et al., 2003). This research has adopted a more qualitative approach, a consideration of which processes contribute to task performance and is seen in many of the studies discuss above. Examples are where phonological representation is explored (e.g. Hitch et al, 1988), attentional control processes are examined (e.g. Cowan et al., 2011) or the contribution of retrieval from secondary memory is considered (e.g. Hamilton et al., 2003).

A further qualitative approach to investigating short term visual memory has been adopted which emphasises the nature, rather than the quantity, of the representation. An approach which assesses the quality, high fidelity and fine detail maintained in the representation (Hamilton, 2011). The protocol may involve a change detection paradigm where the memory stimulus is initially presented then a subsequent stimulus only differs by a very small degree. The participant has to judge whether a change has occurred or not. In the Thompson, Hamilton et al. (2006) study, the Size JND procedure employed such a protocol. The memory stimulus was a single yellow square, at test, another yellow square was presented during retrieval and this stimulus this could be either the same size or a different size; however the size difference could vary in steps from 40% to 5%. In order to detect a change of 5%-10% the participant would have to maintain a high fidelity representation of the initial memory stimulus. In the Thompson et al.study, a delay of 4 seconds between memory and test item led to a performance level of 15%, i.e. for most participants, a smaller change of 5% or 10% was beyond the quality of their representation. Such high fidelity representation was also explored by Dean (2007), Dent (2010), and McConnell and Quinn (2004) with texture, colour and size attributes respectively. A consistent observation in these studies has been the impact of Dynamic Visual Noise (DVN) interference upon the performance. This is an interesting observation given that DVN interference does not appear to impact upon VPT performance (e.g. Andrade, Kemps, Werniers, May & Szmalec, 2002).

This qualitative/quantitative distinction has also been employed in the Luck and Vogel (1997) change detection protocol. Thus in a *categorical change* context, an item in the array may change from a face stimulus at encoding to a cube stimulus in the test phase. In contrast, in a *within-category* change context, where high fidelity representation is required, a memory item such as a face may at test be another face, thus requiring high quality representation of the original face to be remembered (Scolari, Vogel, Awh, 2008). The use of such protocols has led to a debate about capacity in VSWM and the extent to which the need for high fidelity representation may constrain the number of integrated objects held within working memory (see Alvarez & Cavanagh, 2004; Awh, Barton & Vogel, 2007; Bays, Gatalo & Husain, 2009). However, for the purposes of the present chapter, it is interesting to note that this qualitative approach, assessing the high fidelity quality of visual memory, has been employed in a developmental context.

Figure 6 below indicates recent observations by Bull, Hamilton and Pearson (2007) which looked at developmental changes across two child age groups (5-8, 9-12) and an adult group in three tasks, the Size JND/Colour JND (collapsed score), VPT and Corsi task. The figure indicates a clear pattern of developmental fractionation, with significantly less age associated change in the JND task performance than in either the VPT or Corsi tasks. This suggests that in the younger children, performance in this task is relatively mature. A recent study has further examined the quality of representation in children. Heyes, Zokaei, van der Staaij, Bays and Husain (2012) adopted their adult protocol for use with children. This study looked at changes in the precision of memory for orientations, with either 1 bar or 3 bar (sequential presentation) stimulus conditions. The procedure did not employ a change detection protocol, at retrieval a single probe identified the bar to match and participants employed a rotating dial to match the bar with the cued memory item orientation. The quality of representation was measured by identifying the discrepancy in orientation between the memory and rotated test item. This study revealed a number of interesting observations.

Precision of the representation increased from 9 to 13–years of age, both in the 1–item and 3-item conditions; however the 1-item fidelity was much greater than in the 3-item context (Heyes et al, Figure 2, p533). However, given that the 3-item condition involved sequential presentation, and a consideration of the issues in sequential presentation which were identified above, a qualification of the interpretation may be needed. However, the authors present the SPC data (Figure 3) and a strong single item recency effect is seen, much like the Hitch et al, (1988) and the Phillips and Christie, (1977), pattern of results. Thus, a collapsed 3-item figure perhaps obscures the more informative pattern. The authors interpreted their findings as support for a single resource model where the demands for high resolution demands will impinge upon the number of discrete chunks or objects which can be maintained within VSWM. One potential interpretation of the recency item effect in these studies is that the most recent item is under a highest level of activation within a narrow focus of attention able to maintain high resolution of the image, and the pre-recency items under a broader focus of attention where high fidelity cannot be maintained but still at higher than chance levels (Oberauer & Hein, 2011).

Hamilton (2011) has argued that tasks such as the Heyes et al. (2012) task which require high resolution, high fidelity representation, may require cognitive resources in the form of visual attention or *visualization* which are exogenous to working memory processes. He has argued, on the basis of recent research, that the maintenance of high fidelity representation is a consequence of top down visual attention processes acting on a derivative of perceptually driven activity in visual cortex (Berryhill, Chein & Olson, 2011; Gazzaley & Nobre, 2011; Serences, Ester, Vogel & Awh, 2009; Sergent, Ruff, Barbot, Driver & Rees, 2011; Todd, Han, Harrison, & Marois, 2011). As such, these high quality representations are unlikely to be supported by either visual or verbal LTM semantics, and thus have no (stable) place in the working memory workspace which is reserved for objects already bound and integrated by LTM interpretation (Cowan, 1988; Logie, 1995). Evidence from converging cognitive paradigms was employed to support this distinction. DVN impairs high fidelity representations (Dean, 2007; Dent, 2010; McConnell & Quinn, 2004) in the absence of an impact upon a conventional VSWM task, the VPT (Andrade et al., 2002; Hamilton, 2011). In addition, when attentional allocation can be given to VPT maintenance, there is temporal stability, however even when such allocation is allocated to representing high resolution information, the quality of the representation diminishes (Thompson et al, 2006).There are further implications with the use of these tasks within a developmental context.

Should the high fidelity representation be pre-categorical in nature, how does the development account of passive rehearsal followed by active refreshment relate to improvements in task performance (Camos & Barrouillet, 2011, Tam et al., 2010)? It is unlikely that any form of verbal embellishment, and thus verbal rehearsal will underpin performance on these tasks, the nature of these tasks lend themselves to a visual (only) refreshment process. In addition, Heyes et al. (2012) noted a relationship between precision quality and intelligence task performance. This is in contrast to the adult findings reported by Fukuda, Vogel, Mayr and Awh (2010) where a categorical change, i.e. a quantitative measure, is more predictive of fluid intelligence measures. Thus, it may be the case that in a child research context, the attentional processes associated with visualization have more importance in the relationship with wider cognitive abilities.

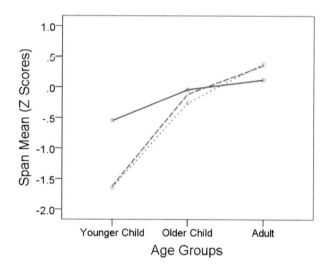

Figure 6. Developmental fractionation in quantitative measures of visuospatial working memory (VPT, ······· Corsi ― ―) and collapsed qualitative measures of short term visual memory (Size/Colour JND ―――).

CONCLUSION

It is clear in the content above that there is incongruence between the visuospatial material employed in a child developmental VSWM research context and the nature of the cognitive resources employed to respond to the task demands. In a task which involves the short term presentation of visuospatial material (on the surface), a complex and diverse range of cognitive resources may be employed, and their deployment and impact appear to be age dependent.

Generic processes clearly exogenous to working memory such as processing speed, impact upon task performance in children, and age related improvements in processing speed may free up cognitive space (e.g. Case, 1982), minimise the decay of information in VSWM by allowing a quicker return to the re-activation of decaying traces (e.g. Towse et al., 1998), or allow greater opportunity for refreshment of the memoranda throughout the procedure (e.g. Barrouillet et al., 2004).

The application of verbal semantics and the increasing use of phonological resources in tasks demanding VSWM also develop through childhood, and this change in strategy may be seen as the hallmark characteristic of child development (Cowan & Alloway, 2009). One would presume that in parallel to the increasing strategic deployment of verbal semantics through childhood, a parallel improvement occurs with visual semantics; such a process would contribute to the enrichment of chunk representation, impacting directly on the compression process (Cowan et al., 2010). The increasing use of strategy and semantics is presumed to make demands upon attentional resources; these controlled processes may impact, during the encoding, maintenance, and retrieval phases of a VSWM task protocol.

Hamilton et al. (2003) argued that in order to disentangle these processes one may employ "...a task-analysis or componential-oriented approach." p66. This is clearly exemplified in the work of Cowan and colleagues (2010, 2011) where discrete attentional control processes at encoding and maintenance were controlled in order to identify developmental changes in the scope of attention or capacity of VSWM. Another way forward is to adopt a qualitative approach and consider visual short term memory for high quality representation, where the usefulness of visual and verbal semantics is necessarily constrained (Bull et al., 2007; Heyes et al., 2012).

However, should working memory in general, and more specifically visuospatial working memory function within the notion of a workspace where available functional architecture scaffolding is critical, then perhaps the *most* important question concerning child development should not address the development of specific cognitive resources, but more, the increasing utilisation of the wider cognitive resources through childhood.

REFERENCES

Alvarez, G. A., & Cavanagh, P. (2004). The Capacity of Visual Short-Term Memory is Set Both by Visual Information Load and by Number of Objects. *Psychological Science, 15*(2), 106-111. doi: 10.1111/j.0963-7214.2004.01502006.x

Andrade, J., Kemps, E., Werniers, Y., May, J., & Szmalec, A. (2002). Insensitivity of visual short-term memory to irrelevant visual information. *The Quarterly journal of experimental psychology. A, Human experimental psychology, 55*(3), 753-774. doi: 10.1080/02724980143000541

Avons, S. E., Ward, G., & Melling, L. (2004). Item and order memory for novel visual patterns assessed by two-choice recognition. [Clinical Trial Randomized Controlled Trial Research Support, Non-U.S. Gov't]. *The Quarterly journal of experimental psychology. A, Human experimental psychology, 57*(5), 865-891. doi: 10.1080/02724980343000521

Awh, E., Barton, B., & Vogel, E. K. (2007). Visual working memory represents a fixed number of items regardless of complexity. [Research Support, N.I.H., Extramural Research Support, U.S. Gov't, Non-P.H.S.]. *Psychological Science, 18*(7), 622-628. doi: 10.1111/j.1467-9280.2007.01949.x

Baddeley, A. (2000). The episodic buffer: a new component of working memory? *Trends in Cognitive Sciences, 4*(11), 417-423. doi: 10.1016/s1364-6613(00)01538-2

Baddeley, A. (2000). The episodic buffer: a new component of working memory? *Trends in Cognitive Sciences, 4*(11), 417-423. doi: Doi 10.1016/S1364-6613(00)01538-2

Baddeley, A. (2012). Working memory: theories, models, and controversies. [Review]. *Annual review of psychology, 63*, 1-29. doi: 10.1146/annurev-psych-120710-100422

Baddeley, A., & Hitch, G. J. (1974). Working Memory. In G. Bower (Ed.), *Recent advances in learning and motivation* (pp. 47-90). New York: Academic Press.

Baddeley, A. D. (1986). *Working Memory*. Oxford: Oxford University Press.

Baddeley, A. D. (2007). *Working memory, thought and action.* Oxford: Oxford University.

Baddeley, A. D., Allen, R. J., & Hitch, G. J. (2011). Binding in visual working memory: the role of the episodic buffer. [Review]. *Neuropsychologia, 49*(6), 1393-1400. doi: 10.1016/j.neuropsychologia.2010.12.042

Baddeley, A. D., & Hitch, G. J. (2000). Development of working memory: Should the Pascual-Leone and the Baddeley and Hitch models be merged? *Journal of experimental child psychology, 77*(2), 128-137. doi: DOI 10.1006/jecp.2000.2592

Barrouillet, P., Bernardin, S., & Camos, V. (2004). Time constraints and resource sharing in adults' working memory spans. [Clinical Trial Randomized Controlled Trial]. *Journal of experimental psychology. General, 133*(1), 83-100. doi: 10.1037/0096-3445.133.1.83

Barrouillet, P., Portrat, S., & Camos, V. (2011). On the law relating processing to storage in working memory. [Meta-Analysis Research Support, Non-U.S. Gov't]. *Psychological Review, 118*(2), 175-192. doi: 10.1037/a0022324

Bays, P. M., Catalao, R. F., & Husain, M. (2009). The precision of visual working memory is set by allocation of a shared resource. [Research Support, Non-U.S. Gov't]. *Journal of vision, 9*(10), 7 1-11. doi: 10.1167/9.10.7

Bays, P. M., Catalao, R. F. G., & Husain, M. (2009). The precision of visual working memory is set by allocation of a shared resource. *Journal of vision, 9*(10), 7-7. doi: 10.1167/9.10.7

Berch, D. B., Krikorian, R., & Huha, E. M. (1998). The Corsi block-tapping task: Methodological and theoretical considerations. *Brain and Cognition, 38*(3), 317-338. doi: DOI 10.1006/brcg.1998.1039

Berryhill, M. E., Chein, J., & Olson, I. R. (2011). At the intersection of attention and memory: the mechanistic role of the posterior parietal lobe in working memory. [Research Support, N.I.H., Extramural Research Support, Non-U.S. Gov't]. *Neuropsychologia, 49*(5), 1306-1315. doi: 10.1016/j.neuropsychologia.2011.02.033

Brown, L. A., Forbes, D., & McConnell, J. (2006). Limiting the use of verbal coding in the Visual Patterns Test. *Quarterly Journal of Experimental Psychology, 59*(7), 1169-1176. doi: 10.1080/17470210600665954

Bull, R., Pearson, D. G., & Hamilton, C. J. (2007). Full Report on 'Measurement and Interference of Visual-Spatial Memory' (Vol. RES 000-22-1517, pp. 13-27): ESRC.

Burnett Heyes, S., Zokaei, N., van der Staaij, I., Bays, P. M., & Husain, M. (2012). Development of visual working memory precision in childhood. [Clinical Trial Research Support, Non-U.S. Gov't]. *Developmental science, 15*(4), 528-539. doi: 10.1111/j.1467-7687.2012.01148.x

Camos, V., & Barrouillet, P. (2011). Developmental change in working memory strategies: from passive maintenance to active refreshing. [Randomized Controlled Trial Research Support, Non-U.S. Gov't]. *Developmental psychology, 47*(3), 898-904. doi: 10.1037/a0023193

Camos, V., & Barrouillet, P. (2011). Developmental Change in Working Memory Strategies: From Passive Maintenance to Active Refreshing. *Developmental psychology, 47*(3), 898-904. doi: Doi 10.1037/A0023193

Camos, V., Mora, G., & Oberauer, K. (2011). Adaptive choice between articulatory rehearsal and attentional refreshing in verbal working memory. [Comparative Study Research Support, Non-U.S. Gov't]. *Memory & cognition, 39*(2), 231-244. doi: 10.3758/s13421-010-0011-x

Case, R., Kurland, D. M., & Goldberg, J. (1982). Operational Efficiency and the Growth of Short-Term-Memory Span. *Journal of experimental child psychology, 33*(3), 386-404. doi: Doi 10.1016/0022-0965(82)90054-6

Cestari, V., Lucidi, A., Pieroni, L., & Rossi-Arnaud, C. (2007). Memory for object location: A span study in children. *Canadian Journal of Experimental Psychology-Revue Canadienne De Psychologie Experimentale, 61*(1), 13-20. doi: Doi 10.1037/Cjep2007002

Conway, A. R., Jarrold, C., Kane, M. J., Miyake, A., & Towse, J. (2007). *Variation in Working Memory*. Oxford: Oxford University Press.

Cornoldi, C., & Vecchi, T. (2003). *Visuo-Spatial Working Memory and Individual Differences*. Hove, UK: Psychology Press.

Cowan, N. (1988). Evolving Conceptions Of Memory Storage, Selective Attention, And Their Mutual Constraints Within The Human Information-Processing System. *Psychological Bulletin, 104*(2), 163-191. doi: 10.1037/0033-2909.104.2.163

Cowan, N. (1999). An embedded-processes model of working memory. In A. Miyake & P. Shah (Eds.), *Models of Working Memory: Mechanisms of active maintenance and executive control*. Cambridge: Cambridge University Press.

Cowan, N. (2011). The focus of attention as observed in visual working memory tasks: Making sense of competing claims. *Neuropsychologia, 49*(6), 1401-1406. doi: 10.1016/j.neuropsychologia.2011.01.035

Cowan, N., & Alloway, T. (2009). Development of Working Memory In Childhood. In M. L. Courage & N. Cowan (Eds.), *The development of memory in infancy and childhood* (2nd ed.). London: Psychology Press.

Cowan, N., AuBuchon, A. M., Gilchrist, A. L., Ricker, T. J., & Saults, J. S. (2011). Age differences in visual working memory capacity: not based on encoding limitations. *Developmental science, 14*(5), 1066-1074. doi: 10.1111/j.1467-7687.2011.01060.x

Cowan, N., Elliott, E. M., Saults, J. S., Morey, C. C., Mattox, S., Hismjatullina, A., & Conway, A. R. A. (2005). On the capacity of attention: Its estimation and its role in working memory and cognitive aptitudes. *Cognitive Psychology, 51*(1), 42-100. doi: DOI 10.1016/j.cogpsych.2004.12.001

Cowan, N., Fristoe, N. M., Elliott, E. M., Brunner, R. P., & Saults, J. S. (2006). Scope of attention, control of attention, and intelligence in children and adults. *Memory & cognition, 34*(8), 1754-1768. doi: Doi 10.3758/Bf03195936

Cowan, N., Hismjatullina, A., AuBuchon, A. M., Saults, J. S., Horton, N., Leadbitter, K., & Towse, J. (2010). With development, list recall includes more chunks, not just larger ones. [Research Support, N.I.H., Extramural]. *Developmental psychology, 46*(5), 1119-1131. doi: 10.1037/a0020618

Dean, G. M., Dewhurst, S. A., & Whittaker, A. (2008). Dynamic visual noise interferes with storage in visual working memory. *Experimental psychology, 55*(4), 283-289. doi: Doi 10.1027/1618-3169.55.4.283

Della Sala, S., Gray, C., Baddeley, A., Allamano, N., & Wilson, L. (1999). Pattern span: a tool for unwelding visuo-spatial memory. *Neuropsychologia, 37*, 1189-1199.

Dent, K. (2010). Dynamic visual noise affects visual short-term memory for surface color, but not spatial location. *Experimental psychology, 57*(1), 17-26. doi: 10.1027/1618-3169/a000003

Engle, R. W., Tuholski, S. W., Laughlin, J. E., & Conway, A. R. A. (1999). Working memory, short term memory, and general fluid intelligence. *Journal of Experimental Psychology: General, 125*, 309-331.

Fisk, J. E., & Sharp, C. A. (2003). The role of the executive system in visuo-spatial memory functioning. *Brain and Cognition, 52*(3), 364-381. doi: 10.1016/s0278-2626(03)00183-0

Friedman, N. P., & Miyake, A. (2004). The reading span test and its predictive power for reading comprehension ability. *Journal of Memory and Language, 51*(1), 136-158. doi: 10.1016/j.jml.2004.03.008

Fry, A. F., & Hale, S. (1996). Processing speed, working memory, and fluid intelligence: Evidence for a developmental cascade. *Psychological Science, 7*(4), 237-241.

Fukuda, K., Vogel, E., Mayr, U., & Awh, E. (2010). Quantity, not quality: the relationship between fluid intelligence and working memory capacity. [Research Support, N.I.H., Extramural]. *Psychonomic bulletin & review, 17*(5), 673-679. doi: 10.3758/17.5.673

Gaillard, V., Barrouillet, P., Jarrold, C., & Camos, V. (2011). Developmental differences in working memory: Where do they come from? [Research Support, Non-U.S. Gov't]. *Journal of experimental child psychology, 110*(3), 469-479. doi: 10.1016/j.jecp.2011.05.004

Gazzaley, A., & Nobre, A. C. (2012). Top-down modulation: bridging selective attention and working memory. *Trends in Cognitive Sciences, 16*(2), 129-135. doi: 10.1016/j.tics.2011.11.014

Hamilton, C., Coates, R., & Heffernan, T. (2003). What develops in visuo-spatial working memory development? *European Journal of Cognitive Psychology, 15*(1), 43-69. doi: 10.1080/09541440303597

Hamilton, C. J. (2011). The nature of visuospatial representation within working memory. In A. Vandierendonck & A. Smalec (Eds.), *Spatial Working Memory* (pp. 122-144). Hove: Psychology Press.

Hasher, L., Lustig, C., & Zachs, R. T. (2007). Inhibitory mechanisms and the control of attention. In A. R. Conway, C. Jarrold, M. J. Kane, A. Miyake & J. Towse (Eds.), *Variation in working memory*. Oxford: Oxford University.

Henry, L. A., Messer, D., Luger-Klein, S., & Crane, L. (2012). Phonological, visual, and semantic coding strategies and children's short-term picture memory span. *Quarterly Journal of Experimental Psychology, 65*(10), 2033-2053. doi: 10.1080/17470218.2012.672997

Hitch, G. J., Halliday, S., Schaafstal, A. M., & Schraagen, J. M. C. (1988). Visual Working Memory in Young-Children. *Memory & cognition, 16*(2), 120-132. doi: Doi 10.3758/Bf03213479

Hitch, G. J., Towse, J. N., & Hutton, U. (2001). What limits children's working memory span? Theoretical accounts and applications for scholastic development. *Journal of Experimental Psychology-General, 130*(2), 184-198.

Hitch, G. J., Woodin, M. E., & Baker, S. (1989). Visual and Phonological Components of Working Memory in Children. *Memory & cognition, 17*(2), 175-185. doi: Doi 10.3758/Bf03197067

Isaacs, E. B., & Varghakhadem, F. (1989). Differential Course of Development of Spatial and Verbal Memory Span - a Normative Study. *British Journal of Developmental Psychology, 7*, 377-380.

Jones, D., Farrand, P., Stuart, G., & Morris, N. (1995). Functional Equivalence of Verbal and Spatial Information in Serial Short-Term-Memory. *Journal of Experimental Psychology-Learning Memory and Cognition, 21*(4), 1008-1018. doi: Doi 10.1037/0278-7393.21.4.1008

Kahneman, D. (1973). *Attention and Effort*. Englewood Cliffs, New Jersey: Prentice-Hall Inc.

Kail, R., & Salthouse, T. A. (1994). Processing Speed as a Mental-Capacity. *Acta Psychologica, 86*(2-3), 199-225.

Kemps, E. (2001). Complexity effects in visuo-spatial working memory: implications for the role of long-term memory. *Memory, 9*(1), 13-27. doi: 10.1080/09658210042000012

Logie, R. H. (1995). *Visuo-Spatial Working Memory*. Hove, Uk: Lawrence Earlbaum Associates Ltd.

Logie, R. H. (1996). The seven ages of working memory. In J. T. E. Richardson, R. W. Engle, L. Hasher, R. H. Logie, E. R. Stoltzfus & R. T. Zachs (Eds.), *Working Memory and Human Cognition*. Oxford: Oxford University Press.

Logie, R. H. (2011). The Functional Organization and Capacity Limits of Working Memory. *Current Directions in Psychological Science, 20*(4), 240-245. doi: 10.1177/0963721411415340

Logie, R. H., & Pearson, D. G. (1997). The inner eye and the inner scribe of visuo-spatial working memory: Evidence from developmental fractionation. *European Journal of Cognitive Psychology, 9*(3), 241-257.

Luck, S. J., & Vogel, E. K. (1997). The capacity of visual working memory for features and conjunctions. *Nature, 390*(6657), 279-281.

Mammarella, N., Pazzaglia, F., & Cornoldi, C. (2006). The assessment of imagery and spatial functions in children and adults. In T. Vecchi & G. Bottini (Eds.), *The assessment of imagery and visuo-spatial working memory in children and adults* (pp. 15-38). Amsterdam: John Benjamins.

Miles, C., Morgan, M. J., Milne, A. B., & Morris, E. D. M. (1996). Developmental and individual differences in visual memory span. *Current Psychology, 15*(1), 53-67. doi: Doi 10.1007/Bf02686934

Milner, B. (1972). Disorders of learning and memory after temporal lobe lesions in man. *Clinical neurosurgery, 19*, 421-446.

Morey, C. C., & Cowan, N. (2004). When visual and verbal memories compete: Evidence of cross-domain limits in working memory. *Psychonomic bulletin & review, 11*(2), 296-301. doi: Doi 10.3758/Bf03196573

Nobre, A. C., & Stokes, M. G. (2011). Attention and short-term memory: Crossroads. *Neuropsychologia, 49*(6), 1391-1392. doi: 10.1016/j.neuropsychologia.2011.04.014

Oberauer, K., & Hein, L. (2012). Attention to Information in Working Memory. *Current Directions in Psychological Science, 21*(3), 164-169. doi: 10.1177/0963721412444727

Orsini, A., Grossi, D., Capitani, E., Laiacona, M., Papagno, C., & Vallar, G. (1987). Verbal and Spatial Immediate Memory Span - Normative Data from 1355 Adults and 1112 Children. *Italian Journal of Neurological Sciences, 8*(6), 539-548.

Palmer, S. (2000). Working memory: a developmental study of phonological recoding. *Memory, 8*(3), 179-193. doi: 10.1080/096582100387597

Pashler, H. (1988). Familiarity and Visual Change Detection. *Perception & Psychophysics, 44*(4), 369-378. doi: Doi 10.3758/Bf03210419

Pentland, L. M., Anderson, V. A., Dye, S., & Wood, S. J. (2003). The Nine Box Maze Test: A measure of spatial memory development in children. *Brain and Cognition, 52*(2), 144-154. doi: 10.1016/s0278-2626(03)00079-4

Phillips, W. A., & Baddeley, A. D. (1971). Reaction Time and Short-Term Visual Memory. *Psychonomic Science, 22*(2), 73-74.

Phillips, W. A., & Christie, D. F. (1977). Interference with visualization. *The Quarterly journal of experimental psychology, 29*(4), 637-650. doi: 10.1080/14640747708400638

Phillips, W. A., & Christie, D. F. M. (1977). COMPONENTS OF VISUAL MEMORY. *Quarterly Journal of Experimental Psychology, 29*(FEB), 117-133. doi: 10.1080 /0033 5557743000080

Pickering, S. J., Gathercole, S. E., Hall, M., & Lloyd, S. A. (2001). Development of memory for pattern and path: Further evidence for the fractionation of visuo-spatial memory. *The Quarterly Journal of Experimental Psychology A, 54*(2), 397-420. doi: 10.1080/ 02724980042000174

Pieroni, L., Rossi-Arnaud, C., & Baddeley, A. (2011). What can symmetry tell us about working memory? In A. Vandierendonck & A. Szmalec (Eds.), *Spatial Working Memory* (pp. 145-158). Hove, UK: Psychology Press.

Portrat, S., Barrouillet, P., & Camos, V. (2008). Time-Related Decay or Interference-Based Forgetting in Working Memory? *Journal of Experimental Psychology-Learning Memory and Cognition, 34*(6), 1561-1564. doi: Doi 10.1037/A0013356

Postma, A., & DeHaan, E. H. F. (1996). What was where? Memory for object locations. *Quarterly Journal of Experimental Psychology Section a-Human Experimental Psychology, 49*(1), 178-199. doi: Doi 10.1080/027249896392856

Quinn, J. G., & McConnell, J. (2004). Cognitive mechanisms of visual memories and visual images. *Imagin Cogn Pers, 23*, 201-207.

Riggs, K. J., McTaggart, J., Simpson, A., & Freeman, R. P. (2006). Changes in the capacity of visual working memory in 5- to 10-year-olds. *Journal of experimental child psychology, 95*(1), 18-26. doi: 10.1016/j.jecp.2006.03.009

Riggs, K. J., Simpson, A., & Potts, T. (2011). The development of visual short-term memory for multifeature items during middle childhood. *Journal of experimental child psychology, 108*(4), 802-809. doi: 10.1016/j.jecp.2010.11.006

Rossi-Arnaud, C., Pieroni, L., & Baddeley, A. (2006). Symmetry and binding in visuo-spatial working memory. *Neuroscience, 139*(1), 393-400. doi: 10.1016/j.neuroscience.2005. 10.048.

Rossi-Arnaud, C., Pieroni, L., Spataro, P., & Baddeley, A. (2012). Working memory and individual differences in the encoding of vertical, horizontal and diagonal symmetry. [Research Support, Non-U.S. Gov't]. *Acta Psychologica, 141*(1), 122-132. doi: 10.1016/j.actpsy.2012.06.007.

Rudkin, S. J., Pearson, D. G., & Logie, R. H. (2007). Executive processes in visual and spatial working memory tasks. *Quarterly Journal of Experimental Psychology, 60*(1), 79-100. doi: 10.1080/17470210600587976.

Scolari, M., Vogel, E. K., & Awh, E. (2008). Perceptual expertise enhances the resolution but not the number of representations in working memory. *Psychonomic bulletin & review, 15*(1), 215-222. doi: 10.3758/pbr.15.1.215.

Sergent, C., Ruff, C. C., Barbot, A., Driver, J., & Rees, G. (2011). Top-Down Modulation of Human Early Visual Cortex after Stimulus Offset Supports Successful Postcued Report. *Journal of Cognitive Neuroscience, 23*(8), 1921-1934. doi: 10.1162/jocn.2010.21553.

St Clair-Thompson, H., & Sykes, S. (2010). Scoring methods and the predictive ability of working memory tasks. [Research Support, Non-U.S. Gov't]. *Behavior research methods, 42*(4), 969-975. doi: 10.3758/BRM.42.4.969.

Sun, H., Zimmer, H. D., & Fu, X. (2011). The influence of expertise and of physical complexity on visual short-term memory consolidation. [Research Support, Non-U.S. Gov't]. *Quarterly Journal of Experimental Psychology, 64*(4), 707-729. doi: 10.1080/17470218.2010.511238.

Tam, H., Jarrold, C., Baddeley, A. D., & Sabatos-DeVito, M. (2010). The development of memory maintenance: children's use of phonological rehearsal and attentional refreshment in working memory tasks. [Research Support, Non-U.S. Gov't]. *Journal of experimental child psychology, 107*(3), 306-324. doi: 10.1016/j.jecp.2010.05.006

Thompson, J. M., Hamilton, C. J., Gray, J. M., Quinn, J. G., Mackin, P., Young, A. H., & Ferrier, I. N. (2006). Executive and visuospatial sketchpad resources in euthymic bipolar disorder: Implications for visuospatial working memory architecture. *Memory, 14*(4), 437-451. doi: 10.1080/09658210500464293.

Thompson, J. M., Hamilton, C. J., Gray, J. M., Quinn, J. G., Mackin, P., Young, A. H., & Nicol Ferrier, I. (2006). Executive and visuospatial sketchpad resources in euthymic bipolar disorder: Implications for visuospatial working memory architecture. *Memory, 14*(4), 437-451. doi: 10.1080/09658210500464293.

Todd, J. J., Han, S. W., Harrison, S., & Marois, R. (2011). The neural correlates of visual working memory encoding: a time-resolved fMRI study. [Research Support, N.I.H., Extramural Research Support, U.S. Gov't, Non-P.H.S.]. *Neuropsychologia, 49*(6), 1527-1536. doi: 10.1016/j.neuropsychologia.2011.01.040.

Towse, J. N., Hitch, G. J., & Hutton, U. (1998). A reevaluation of working memory capacity in children. *Journal of Memory and Language, 39*(2), 195-217.

Unsworth, N., & Engle, R. W. (2007). The nature of individual differences in working memory capacity: Active maintenance in primary memory and controlled search from secondary memory. *Psychological Review, 114*(1), 104-132. doi: 10.1037/0033-295x.114.1.104.

Vandierendonck, A., Kemps, E., Fastame, M. C., & Szmalec, A. (2004). Working memory components of the Corsi blocks task. *British Journal of Psychology, 95*, 57-79. doi: 10.1348/000712604322779460.

Vergauwe, E., Barrouillet, P., & Camos, V. (2009). Visual and Spatial Working Memory Are Not That Dissociated After All: A Time-Based Resource-Sharing Account. *Journal of Experimental Psychology-Learning Memory and Cognition, 35*(4), 1012-1028. doi: 10.1037/a0015859.

Wheeler, M. E., & Treisman, A. M. (2002). Binding in short-term visual memory. *Journal of Experimental Psychology: General, 131*(1), 48-64. doi: 10.1037//0096-3445.131.1.48.

Wilson, J. T. L., Scott, J. H., & Power, K. G. (1987). Developmental Differences in the Span of Visual Memory for Pattern. *British Journal of Developmental Psychology, 5*, 249-255.

Chapter 6

AGING AND VISUAL FEATURE BINDING IN WORKING MEMORY

Richard J. Allen[1], Louise A. Brown[2] and Elaine Niven[3]
[1]Institute of Psychological Sciences, University of Leeds, Leeds, UK
[2]Division of Psychology, Nottingham Trent University, Nottingham, UK
[3]University of Edinburgh, Edinburgh, UK

ABSTRACT

This chapter examines the question of how the features or elements (e.g. shape, color, or spatial location) of a stimulus that is encountered in the visual environment are bound together in working memory to form an integrated representation. We first briefly review recent research on this topic in healthy young adults, examining the factors that determine successful encoding and retention in working memory. A particular focus of the chapter concerns how this key cognitive process (or set of processes) might vary as a function of healthy cognitive aging, and of neuropsychological disorders typically associated with aging (e.g. Alzheimer's disease). It appears that while older adults typically show associative deficits in long-term memory, age-related binding deficits in working memory are somewhat inconsistent in nature, and may depend to some extent on the form of binding being examined.

INTRODUCTION

Previous research on working memory (WM) has often focused on how we process, store and retrieve individual elements, such as form, color, size, or location (in visuospatial memory), and meaning or phonological form (in auditory-verbal memory). In a similar vein, a large majority of work has examined memory for information encountered within a single modality (e.g. vision or audition). This has developed at least partly through efforts to retain simplicity of design and interpretation. As such, classic models of WM (e.g. Baddeley and Hitch, 1974; Logie, 1995) focused on capturing distinctions between elements and between modalities. However, it is clearly also important to explore processing of more complex stimuli, including items and episodes that consist of multiple conjunctions of elements, both within and between modalities. This is often how information is encountered and indeed consciously experienced in the real world. For example, items in the visual environment typically vary along multiple dimensions such as shape and color; how do we not only perceive but also remember that the apple was red and the leaf green, rather than vice versa? This issue of information binding is a broad one, applying across contexts, domains, and forms of information, as well as having precise and meaningful applications. For example, how does a patient on multiple courses of medication remember, for example, the shape and color of the pill they have taken? By examining such questions, we can develop more sophisticated and comprehensive models of function and structure. Furthermore, we can utilize the methods and theoretical insights that result in understanding how memory changes as a result of both typical cognitive development and of brain disorders, with the potential aim of developing assessment and rehabilitation tools to assist in identifying and ameliorating associated problems. This chapter will focus for the most part on binding within visuospatial WM, briefly considering some of the key developments on this topic from research on young adults, before examining how binding and relational memory might be characterized in LTM and WM as a function of healthy cognitive aging, and Alzheimer's Disease (AD).

VISUOSPATIAL BINDING IN WORKING MEMORY

Binding in WM has received considerable attention in the last few years. This is partly the result of efforts to resolve the shortcomings of influential WM models in addressing how different forms of binding may operate. For example, the tripartite model described by Baddeley and Hitch (1974) was successful in setting out how simple verbal or visuospatial information is temporarily stored (in the phonological loop and visuospatial sketchpad subcomponents respectively) and manipulated (via the central executive, an attentional control resource). However, Baddeley (2000) noted that this simple model was not able to adequately capture how constituent information is bound into objects within or between modalities, or how this is connected to existing representations in long-term memory. As a result, Baddeley added a fourth component to WM termed the episodic buffer, described as a modality-independent store responsible for binding and acting as the central hub between incoming information and LTM. In order to retain parsimony and generate testable predictions, it was originally assumed that binding within the episodic buffer was an active and effortful process, requiring central executive control. Thus, remembering the conjunctions between shape and color, for example, should be more dependent on attention than simply remembering the individual elements.

In order to test this and related questions, researchers have often adopted variants of a recognition paradigm. This involves an initial presentation of a to-be-remembered display containing target stimuli (e.g. a set of colored shapes), followed (after a short delay) by a test phase. This can be a repetition of the whole display, with participants required to detect whether a change has occurred (e.g. Phillips, 1974; Luck and Vogel, 1997), or a single recognition probe, in which case participants must decide whether this was present in the original display (e.g. Wheeler and Treisman, 2002). Memory for individual features is assessed through judgments of whether a new feature is present at test. This test method has also sometimes been used as an implicit measure of memory for binding (e.g. Elsley and Parmentier, 2009; Luck and Vogel, 1997), though a more direct method of tapping binding involves judgments of whether specific test combinations of color and shape were part of the original display (with lure trials involving recombined objects (e.g. Allen, Baddeley and Hitch, 2006; Wheeler and Treisman, 2002). A minority of research has also gone beyond recognition paradigms, using tools such as cued verbal recall and non-verbal reconstruction (e.g. Pertsov, Dong, Peich, and Husain, 2012; Ueno, Mate, Allen, Hitch, and Baddeley, 2011).

Generally using single probe recognition, research to date has failed to consistently support the notion of an attention-demanding binding process in healthy adult participants; adding a demanding concurrent task such as backward counting in decrements of three (aimed at reducing available executive resources) during stimulus presentation has substantial effects on visual memory performance, but does not appear to reliably disrupt shape-color binding any more than memory for the individual features (Allen et al. 2006; Allen, Hitch, Mate, and Baddeley, 2012; but see Brown and Brockmole, 2010). This apparent lack of requirement for attention in encoding bound information has also been extended to contexts in which shape and color are initially presented separately in space or time (Karlsen, Allen, Baddeley, and Hitch, 2010) or across visual and auditory modalities (Allen, Hitch, and Baddeley, 2009). These findings led Baddeley, Allen, and Hitch (2011) to suggest that, at least for simple feature conjunctions encountered in the environment, binding is likely a perceptual process that proceeds relatively automatically, with the episodic buffer (or similar component) a passive recipient of this processing. The constraints that might be applied to this automaticity are yet to be clearly mapped out, however; it may be that binding, in certain forms and contexts, is in fact particularly effortful. For example, Ecker, Maybery, and Zimmer (2013) have recently argued that, while intrinsic binding (between elements that are part of the same perceptual unit or 'object') may proceed relatively automatically, extrinsic binding (between different contextual elements, that are part of the same episode but not the same stimulus) may require additional cognitive resources.

It is also important to consider how features and their conjunctions are retained in memory, once encoded. Luck and Vogel (1997; Vogel, Woodman, and Luck, 2001) found evidence to indicate that visual short-term memory is capacity-limited in terms of number of objects rather than elements, at around 4 items in size. This was challenged by Wheeler and Treisman (2002), who found that memory for feature binding was worse when recognition memory was tested by re-presenting the whole display again, rather than using a single test probe. They instead claimed that features such as form, color or location are retained in separate stores, and that the bindings between them are reliant on attention to stay intact. While some evidence using concurrent load methodology supports this (e.g. Brown and Brockmole, 2010; Fougnie and Marois, 2009), the majority of findings using this and similar approaches indicate that maintenance of bindings, in healthy young adults, is not any more dependent on attention than is the temporary retention of visual information in general (e.g. Delvenne, Cleeremans, and Laloyaux, 2010; Gajewski and Brockmole, 2006; Johnson, Hollingworth, and Luck, 2008; Morey and Bieler, 2013; Yeh, Yang, and Chiu, 2005).

While encoding and maintenance of simple feature bindings may be relatively cost-free in terms of general attentional resources, it is emerging that bound representations are potentially fragile and susceptible to interference caused by subsequently encountered stimuli. For example, when colored shapes are presented serially (as opposed to simultaneously, as in the research discussed so far), memory for the bindings between these features is disproportionately worse, particularly at early positions in the sequence (Allen et al., 2006; Brown and Brockmole, 2010). In the same way, presenting certain forms of visual 'suffix' (i.e. a further item to-be-ignored) during the retention interval can particularly disrupt binding memory (Ueno et al., 2011a, 2011b). Similarly, using a precision-based approach to memory for color and orientation, Gorgoraptis, Catalao, Bays, and Husain (2011) have recently observed an increase in binding errors when items are presented sequentially, specifically caused by disruption of earlier stored representations rather than by temporal decay. This fragility has also been found to be a key factor in whether participants are able to learn shape-color bindings over multiple trials (Logie, Brockmole, and Vandenbroucke, 2009). Regarding how information is lost from memory, Cowan, Blume and Saults (in press) have recently suggested that features may be bound in memory (with a capacity limit of only around 3 objects), but these representations can be incomplete; it is possible to retain information about an item's color, for example, while forgetting its shape (cf. Gajewski and Brockmole, 2006). Thus, when interference and information loss occurs, it does not necessarily proceed in an 'all-or-none' manner.

It appears that for healthy young adults, while encoding and retention in visuospatial WM is reliant on central executive control (e.g. Miyake et al., 2001; Morey and Cowan, 2005), simple feature binding is not more so, but it is easily lost from memory. It should be noted however that the majority of research on this topic has examined binding between particularly salient surface features such as color and shape. In addition, a focus has also been placed on the binding of 'unitized' features, presented as part of the same object within space. Caution should be exercised when collapsing across different forms of relational and conjunctive processing; we cannot necessarily generalize findings and assumptions to other forms of binding, e.g. between identity and location, between features within the same dimension (e.g. color-color binding), or between features drawn from different objects/locations, all of which may behave differently and draw on different cognitive processes (e.g. Delvenne and Bruyer, 2004; Ecker et al., 2013; Wheeler and Treisman, 2002; Xu, 2002). Nevertheless, recent research has been relatively successful in starting to delineate more clearly how binding may operate in visuospatial working memory. In the same way that we should be aware of how cognitive processes might vary across contexts, it is also important to consider how they may fluctuate across the lifespan.

AGE-RELATED ASSOCIATIVE DEFICITS IN LONG-TERM MEMORY

It has long been known that, in adulthood, episodic memory ability declines in older age. All memorial aspects contributing to episodic memory performance are not, however, uniformly influenced by age. An early meta-analysis by Spencer and Raz (1995) assessed the extent to which aging effects varied with frequently manipulated task demands (for example, recall versus recognition and questions on content versus those on contextual information). Results suggested that memory for content of an episode and memory for the context surrounding and supporting presentation of content are differentially impaired in age comparisons, with memory for context information showing a significantly larger decline with aging. In reviewing the surrounding literature, Chalfonte and Johnson (1996) considered the existing and varied definitions of 'content' and 'context' information, noting that definitions of the latter range from source information (such as modality or speaker of content) to experimenter defined saliency to information not explicitly instructed for memorization.

The study of context deficits (Spencer and Raz, 1995) and use of a source monitoring framework (Johnson, Hashtroudi and Lindsay, 1993) have been extensively employed in exploring episodic memory. However, Chalfonte and Johnson (1996) focused on characterizing age-related impairment in relation to a question applicable across these approaches: whether older adults are able to maintain individual features within a complex memory (such as item identity, color, location, temporal information, source of presented information) and to maintain associations between these component features. In a series of experiments, older and younger adults were presented with a 7 x 7 grid containing 30 colored, identifiable objects (e.g. telephone, bed, spider) in pseudo-random locations, for a minute and a half, with the instruction to remember one specified feature or feature-feature combination. The same time was then given to consider an instruction consistent test array – that is, consisting of either objects, locations, colors, grid-located items or colored items - with the requirement to identify multiple targets from lures. When instructed to remember location information, older adults' feature memory was impaired relative to that of younger adults, and memory for object-location bindings was also significantly poorer than that of younger adults. In contrast, older adults did not exhibit a feature memory deficit when instructed to remember either color or object identity information, yet an age related binding deficit was observed when tested on the combination of these features. The critical finding of older adults' impairment in memory for color-object identity binding, despite corresponding single feature equivalency with younger adults, was replicated when binding test arrays contained lures created via recombination of non-target features. Previously, lures had been created using one feature from initial presentation (e.g. color) and one novel feature (e.g. a new object) thus this finding, specifically, indicated the binding deficit existed due to poorly formed or retrieved associations between features. Moreover, response methods between feature and association tests were equated in this procedure, employing recognition for both, rather than recognition and recall as has been the case in many episodic, context or source monitoring procedures.

Subsequently, much research has focused on binding difficulties as a means of explaining episodic memory failure in older adults. Naveh-Benjamin (2000) posited the associative deficit hypothesis which suggests that older adults are impaired in their ability to form and retrieve links between 'units' of information – units being defined as items, an item's context, contextual elements or "mental codes" (Naveh-Benjamin, 2000, p 1170) in a broad and encompassing manner akin to that of Chalfonte and Johnson (1996). Importantly, this age-related associative, or binding, deficit is characterized as the existence of a memory deficit for the associated (bound) information, over and above any age-related deficit which may exist for the individual (or unbound) units of information. Naveh-Benjamin revealed the existence of such an associative deficit in older adults when testing recognition of items (words and nonwords) and associations between these items (word-nonword pairs or word-pairs) and also when testing recognition of associations of features presented as one item (word, font of presentation and word-font association). The results of a great deal of studies now indicate an age-related binding deficit in tests of long term memory (see Old and Naveh-Benjamin, 2008a, for a meta-analysis). The deficit has been shown across a variety of stimuli including word pairs, picture pairs, and face-name pairs (e.g. Naveh-Benjamin, Guez, Kilb, and Reedy, 2004; Naveh-Benjamin, Hussain, Guez and Bar-On, 2003), face-face and face-location pairs (Bastin and Van der Linden, 2005), and for associations between people and their actions (Old and Naveh-Benjamin, 2008b).

Evidence from studies employing numerous materials suggests that older adults do not appear to form associations available under incidental encoding conditions (where associated units are observed at presentation with instruction only to remember single units for visually presented words), even when such associations can prove advantageous. For example, older adults' recognition of words did not benefit from a rich encoding context, while the performance of younger adults was enhanced in a rich compared to a weak visual context (Fernandes and Manios, 2012). Furthermore, Naveh-Benjamin (2000) investigated incidental versus intentional (instructed) encoding of associations and found support for existing results (Chalfonte and Johnson, 1996) indicating that older adults' associative deficit is even greater under intentional learning conditions compared to incidental learning conditions (see also Old and Naveh-Benjamin, 2008a). Recent work has indicated that older adults' associative deficit may in part arise from poor or ineffectual strategy use at encoding or retrieval and that this deficit may be attenuated or eliminated when effective strategies are stimulated (Naveh-Benjamin and Kilb, 2012; Naveh-Benjamin, Brav and Levi, 2007). Such findings provide an interesting avenue for future studies, however it can be noted that a manipulation of full or divided attention reportedly did not interact with age or the variables affecting strategy use (Naveh-Benjamin and Kilb, 2012), as is consistent with previous work indicating divided attention does not differentially affect binding performance of younger and older adults (Kilb and Naveh-Benjamin, 2007, though see Castle and Craik, 2003). This rejection of attentional control as a key contributor to age differences in association formation is somewhat at odds with the claims by Hasher, Zacks, and May (2009) that older adults have particular problems with inhibitory control. Indeed, Campbell, Hasher, and Thomas (2010) have suggested that age-related associative deficits can actually be attributable to 'hyper-binding' in older adults, with reduced inhibitory control leading to formation of extraneous associations in memory, which in turn negatively impinges on goal-relevant binding.

Despite ongoing investigation of underlying causes, the robust evidence for the presence of a specific binding deficit in LTM has prompted recent attempts to establish whether or not such age related associative deficits exist in the working memory domain.

AGING EFFECTS ON VISUOSPATIAL BINDING IN WORKING MEMORY

Mitchell, Johnson, Raye, Mather, and D'Esposito (2000) investigated the influence of aging on visuospatial binding tasks in working memory by employing a similar paradigm to that used by Chalfonte and Johnson (1996). They assessed memory for everyday objects and their locations, which were presented sequentially (i.e., one item at a time) within a 3 x 3 grid. The different forms of memory (object, location, combination) were tested across discrete blocks of trials. The results showed a specific age-related binding deficit, and that this was driven by an increased false alarm rate, that is, difficulty rejecting lure objects that had been created by recombining features presented in the to-be-remembered array. The authors further showed that the binding deficit was not fully accounted for by the increased test load in the combination trials relative to the individual feature trials. Further work by the same group, using the same paradigm, suggested that impaired hippocampal functioning, and possibly also right prefrontal cortex functioning, underlies the behavioral deficit (Mitchell, Johnson, Raye, and D'Esposito, 2000). The authors concluded that visuospatial binding in working memory might be reliant upon a network in which the prefrontal cortex facilitates hippocampal processing. Indeed, further studies have implicated the hippocampus in visuospatial binding in working memory (Olson, Page, Moore, Chatterjee, and Verfaellie, 2006; Piekema, Kessels, Mars, Petersson, and Fernández, 2006; but see Jeneson and Squire, 2012), though this may depend on the type of binding required; we will return to this later in the chapter.

Cowan, Naveh-Benjamin, Kilb, and Saults (2006) also demonstrated an age-related deficit in memory for colors bound to locations, which was strongest under conditions in which the feature and binding memory trials were mixed within the same block, rather than when trials were presented in discrete blocks according to the form of memory being tested. Cowan et al. also found that the older adults were biased towards responding 'no change', particularly during binding trials. They interpreted this to mean that older adults were relying on familiarity rather than on recollective processes, meaning that they would often overlook the more subtle changes that existed within the array of items when binding memory was being tested. Borg, Leroy, Favre, Laurent, and Thomas-Antérion (2011) also recently demonstrated an age-related deficit for item-location binding. Despite all of this initial evidence of a specific age-related binding deficit in visuospatial working memory, subsequent evidence has been inconsistent. This is because the age-related binding deficit has been fairly elusive when considering the binding of surface visual features such as color and shape, suggesting that the involvement of location as a to-be-remembered feature could be an important factor.

Across three experiments in which both recall and recognition of individual visual features and their combinations was assessed in young and older adults, Brockmole, Parra, Della Sala, and Logie (2008) found that, although older adults exhibited a reliable memory deficit overall, memory for the bindings between these features was not specifically affected by age (see also Parra, Abrahams, Logie, and Della Sala, 2009; and Parra, Abrahams, Fabi, Logie, Luzzi, et al., 2009, for verbal recall of visual material). Furthermore, while Brown and Brockmole (2010) demonstrated an age-related binding deficit in the binding of surface visual features (color and shape), this was in only one of two experiments. In the first experiment working memory for individual colors and shapes, as well as for the combinations, was assessed in young and older adults under conditions of single- (two-digit number repetition) and dual-task (counting backwards in 3s from a two-digit number) conditions. While the results showed a specific binding deficit in attention demanding conditions overall (in contrast to the findings reported by Allen et al., 2006, 2012), there was no specific binding deficit in older adults. In the second experiment the attention manipulation involved comparing simultaneous versus sequential presentation conditions (cf. Allen et al., 2006). As predicted, sequential conditions were disproportionately disruptive for binding memory in both age groups; however, a specific age-related binding deficit was observed overall, independently of the attention manipulation. Age-related binding deficits in specifically visual working memory are therefore observable, but they appear to be unreliable and also quite modest. This conclusion has been supported by recent research we have carried out, in which a small age-related binding deficit was observed in only one of four experiments (Brown, Niven, Logie, and Allen, 2012).

Variation in methodology more generally may also be contributing to the inconsistency in findings within the working memory domain. For example, using the same paradigm to investigate binding in long- as well as short-term memory, Chen and Naveh-Benjamin (2012) reported that the binding of faces with scenes exhibits a specific age-related deficit in both memory domains. However, in relation to the research investigating visual working memory binding (e.g., Brockmole et al., 2008; Brown and Brockmole, 2010), there are contrasts such as in the active nature of the binding that must be carried out at encoding. That is, Chen and Naveh-Benjamin required extrinsic binding while the other studies involved presenting intrinsically bound objects. The stimuli themselves were also more complex, albeit perhaps more ecologically valid (i.e. faces and scenes rather than basic colors and shapes). Furthermore, a continuous recognition paradigm was used by Chen and Naveh-Bejnamin, in which items presented for study and test are intermixed, rather than presenting distinct trials featuring ordered study and test phases. This may be important, particularly as research has shown the fragility of bindings in working memory and the relative ease with which they may be overwritten by subsequent information (Allen et al., 2006; Brown and Brockmole, 2010; Ueno et al., 2011a).

One apparent consistency within the literature on visuospatial binding in working memory is that limited attentional capacity in older adults does not appear either to fully account for (Cowan et al., 2006) or bring about (Brown and Brockmole, 2010) the deficits. This may relate to the argument that the hippocampus is the key relational processor required for binding. Thus, rather than an attention deficit, an impairment in associative processing or memory per se may underlie age-related visuospatial binding deficits in working memory, where they do exist.

VISUOSPATIAL BINDING IN WORKING MEMORY IN ALZHEIMER'S DISEASE

In contrast to the inconsistent findings within the literature on healthy aging, a marked, reliable working memory binding deficit has been observed in AD patients. For example, Parra, Abrahams, Fabi, Logie, Luzzi, et al. (2009) compared working memory for color-shape bindings, verbally recalled, in AD patients with that of healthy, age- and education-matched control participants. Importantly, effects of overall memory capacity were controlled by calibrating the number of to-be-remembered items across the two participant groups, to the point where performance for individual feature memory was equivalent. The authors found that, in the AD patients only, binding performance was poorer than individual feature memory, concluding that AD markedly and specifically affects mechanisms responsible for binding features in working memory (see also Borg et al., 2011; Parra, Abrahams, Logie, and Della Sala, 2010; Parra, Abrahams, Logie, Méndez, Lopera, et al., 2010). Parra, Della Sala, Abrahams, Logie, Méndez, et al. (2011) further showed that a specific binding impairment is also present in AD patients in within-dimension (color-color) binding, and concluded that, rather than relying specifically upon the hippocampus, working memory binding may depend upon surrounding cortices (i.e. perirhinal and entorhinal cortex) that is known to degenerate with AD. Remarkably, Della Sala et al. (2010) demonstrated that specifically AD, and not any of a number of non-AD forms of dementia, is sensitive to working memory binding deficits. They argued that this supports the claim that extrahippocampal regions are critical to working memory binding, due to the earlier and more severe effects seen in these areas with AD as compared with non-AD dementias.

How might this relate to other research indicating a role for the hippocampus in certain forms of age-related binding deficit, and to evidence on WM binding and the hippocampus in general? As already noted, an important distinction might exist between intrinsic and extrinsic, or conjunctive and relational, binding (e.g. Ecker et al., 2013; Moses and Ryan, 2006). It is possible that, for both working memory and LTM, the hippocampus might have a particular role in building and retaining extrinsic or relational bindings, that is, between elements from different objects or parts of an object, and binding to location and other contextual information (Ecker et al., 2013, but see Jeneson and Squire, 2012). In these cases, older adults may show impairment at least partly as a result of hippocampal degradation associated with healthy aging (West, 1993). In contrast, intrinsic or conjunctive binding (between elements that are unitized as part of the same object) may operate independently of the hippocampus, at least for WM. For example, Baddeley, Allen, and Vargha-Khadem (2010) observed intact visual WM for the binding of shape and color in Jon, a patient with developmental amnesia resulting from highly selective hippocampal damage. Thus, WM binding deficits in aging may be attributable to PFC and hippocampal loss connected to extrinsic binding, while further impairment shown by AD patients in intrinsic binding may reflect involvement of and damage to extrahippocampal areas (which are relatively unaffected by cognitive aging). Somewhat in line with this, Parra and colleagues (Parra, Cubelli, and Della Sala, 2011; Parra, Della Sala, Logie and Abrahams, 2009) have reported a case study of an individual with damage to the anterior pole of the left medial temporal lobe, who demonstrated impaired 'intrinsic' shape-color binding in visual WM, alongside intact binding between colors (analogous to extrinsic, or relational binding). This possible intrinsic-extrinsic distinction may cut across WM and LTM, in terms of aging deficits and of the involvement of the hippocampus (though Jeneson and Squire, 2012, have recently argued against any medial temporal lobe contribution to WM). Alternatively, such a distinction may be a key factor only in WM, with any form of binding showing an age-related impairment in long-term memory tasks.

FUTURE DIRECTIONS

As indicated by the research described so far, it remains to be seen to what extent different forms of information binding are consistently intact or impaired as a result of the healthy aging process. Certainly in terms of binding and association formation in LTM, older adults appear to demonstrate reliable deficits relative to individual element retention, and to young adults. Within WM however, this may vary to a greater extent, possibly depending on the type of stimulus and conjunction involved. It is important that future research attempts to address more precisely the boundary conditions at which older adults consistently show intact or impaired performance. As current work might indicate (e.g. Brown et al., 2013), this is not necessarily a straightforward process, as it is possible to observe occasional but inconsistent patterns of impairment in older participants. Nevertheless, identifying the conditions under which reliable deficits emerge remains a viable aim. Thus, examination of factors such as within- vs. cross-modality binding, unitization, and the intrinsic-extrinsic distinction already discussed, would be potentially fruitful directions to pursue in future. Examining how different aspects of WM and attention might be involved in age-related binding deficits would be of value, as different sub-components may be variably affected by cognitive aging (e.g. Craik, Morris, and Gick, 1990; Reuter-Lorenz and Jonides, 2008). Similarly, exploring the interface between WM and LTM in this context would be useful, in order to potentially identify over what timeframes binding deficits become more problematic with age, and how different performance patterns over the short- and long-term can be reconciled. Finally, a greater consideration of the role of strategic processing would be important; the extent to which older adults are able to implement strategies to support memorial performance, and how this relates to any patterns of impairment that are observed, may help clarify when binding itself is genuinely and reliably impaired, and when this is a result of either more general deficits in strategy-use, or idiosyncratic strategy use obscuring real deficits.

Conclusion

Recent developments have begun to elucidate how feature binding might operate, with implications for the possible structure of working memory, interactions with attention, and what determines information loss over short delays and in the face of interference. Different cognitive processes and brain regions might contribute to binding, depending on factors such as the nature of the constituent features and the binding that is required, their relative spatial configurations, and the type of memory task involved. This may be important in determining whether binding declines as a function of cognitive aging. A number of studies have observed such a decline, while others have failed to observe particular binding deficits above and beyond any problems with memory for individual features. It is important that subsequent research attempts to clearly delineate the conditions that consistently give rise to age-related binding deficits. While it remains a challenge to draw this out amid the greater inconsistency and variability often associated with older adults' performance, doing so will not only aid understanding of the cognitive aging process, but also of disorders such as Alzheimer's disease.

References

Allen, R. J., Baddeley, A. D., and Hitch, G. J. (2006). Is the binding of visual features in working memory resource-demanding? *Journal of Experimental Psychology: General, 135,* 298-313.

Allen, R. J., Hitch, G. J., and Baddeley, A. D. (2009). Cross-modal binding and working memory. *Visual Cognition, 17,* 83-102.

Allen, R.J. Hitch, G.J., Mate, J., and Baddeley, A.D. (2012). Feature binding and attention in working memory: A resolution of previous contradictory findings. *Quarterly Journal of Experimental Psychology, 65 (12),* 2369-2383.

Baddeley, A. D. (2000). The episodic buffer: A new component of working memory? *Trends in Cognitive Sciences, 4*, 417-423.

Baddeley, A.D., Allen, R.J., and Hitch, G.J. (2011). Binding in visual working memory: The role of the episodic buffer. *Neuropsychologia, 49,* 1393-1400.

Baddeley, A. D., Allen, R. J., and Vargha-Khadem, F. (2010). Is the hippocampus necessary for visual and verbal binding in working memory? *Neuropsychologia, 48,* 1089-1095.

Baddeley, A. D., and Hitch, G. J. (1974). Working memory. In G. A. Bower (Ed.), *The psychology of learning and motivation: Advances in research and theory.* (Vol. 8, pp. 47-89). New York: Academic Press.

Bastin, C. and Van der Linden, M. (2005) The effects of aging on the recognition of different types of associations. *Experimental Aging Research, 32,* 61-77.

Borg, C., Leroy, N., Favre, E., Laurent, B., and Thomas-Antérion, C. (2011). How emotional pictures influence visuospatial binding in short-term memory in ageing and Alzheimer's disease? *Brain and Cognition, 76,* 20-25.

Brockmole, J. R., Parra, M. A., Della Sala, S., and Logie, R. H. (2008). Do binding deficits account for age-related decline in visual working memory? *Psychonomic Bulletin and Review, 15,* 543-547.

Brown, L. A., and Brockmole, J. R. (2010). The role of attention in binding visual features in working memory: Evidence from cognitive ageing. *Quarterly Journal of Experimental Psychology, 63,* 2067-2079.

Brown, L. A., Niven, E. H., Logie, R. H., and Allen, R. J. (2013). *Encoding and interference effects on visual working memory binding in young and older adults.* Manuscript in preparation.

Campbell, K.L., Hasher, L., and Thomas, R.C. (2010). Hyper-binding: A unique age effect. *Psychological Science, 21 (3),* 399-405.

Castle, A. D. and Craik, F. I. M. (2003). The effects of aging and divided attention on memory for item and associative information. *Psychology and Aging, 18(4),* 873-885.

Chalfonte, B. L., and Johnson, M. K. (1996). Feature memory and binding in young and older adults. *Memory and Cognition, 24,* 403-416.

Cowan, N., Blume, C.L., and Saults, J.S. (in press). Attention to attributes and objects in working memory. *Journal of Experimental Psychology: Learning, Memory, and Cognition.*

Cowan, N., Naveh-Benjamin, M., Kilb, A., and Saults, J. S. (2006). Life-span development of visual working memory: When is feature-binding difficult? *Developmental Psychology, 42,* 1089-1102.

Craik, F.I.M., Morris, R.G., and Gick, M.L. (1990). Adult age differences in working memory. In G. Vallar and T. Shallice (Eds.). *Neuropsychological impairments of short-term memory* (pp. 247-267). New York: Cambridge University Press.

Della Sala, S., Parra, M. A., Fabi, K., Luzzi, S., and Abrahams, S. (2012). Short-term memory binding is impaired in AD but not in non-AD dementias. *Neuropsychologia, 50,* 833-840.

Delvenne, J.-F., and Bruyer, R. (2004). Does visual short-term memory store bound features? *Visual Cognition, 11,* 1-27.

Delvenne, J.-F., Cleeremans, A., and Laloyaux, C. (2010). Feature bindings are maintained in visual short-term memory without sustained focused attention, *Experimental Psychology, 57 (2),* 108-116.

Ecker, U.K.H., Maybery, M., and Zimmer, H.D. (2013). Binding of intrinsic and extrinsic features in working memory. *Journal of Experimental Psychology: General, 142*, 218-234.

Elsley, J., and Parmentier, F. B. R. (2009). Is verbal-spatial binding in working memory impaired by a concurrent memory load? *Quarterly Journal of Experimental Psychology, 62*, 1696-1705.

Fernandes, M. A. and Manios, M. (2012). How does encoding context affect memory in younger and older adults? *The Quarterly Journal of Experimental Psychology, 65(9)*, 1699-1720.

Fougnie, D., and Marois, R. (2009). Attentive tracking disrupts feature binding in visual working memory. *Visual Cognition, 17*, 48-66.

Gajewski, D. A., and Brockmole, J. R. (2006). Feature bindings endure without attention: Evidence from an explicit recall task. *Psychonomic Bulletin and Review, 13*, 581-587.

Gorgoraptis, N., Catalao, R.F.G., Bays, P.M., and Musain, M. (2011). Dynamic updating or working memory resources for visual objects. *The Journal of Neuroscience, 31 (23)*, 8502-8511.

Hasher, L., Zacks, R.T., and May, C.P. (1999). Inhibitory control, circadian arousal, and age. In D. Gopher and A. Koriat (Eds.), *Attention and performance XVII* (pp. 653-675). Cambridge, MA: MIT Press.

Jeneson, A., and Squire, L.R. (2012). Working memory, long-term memory, and medial temporal lobe function. *Learning and Memory, 19*, 15-25.

Johnson, M. K., Hashtroudi, S., and Lindsay, D. S. (1993) Source monitoring. *Psychological Bulletin, 14(1)*, 3-28.

Johnson, J. S., Hollingworth, A., and Luck, S. J. (2008). The role of attention in the maintenance of feature bindings in visual short-term memory. *Journal of Experimental Psychology: Human Perception and Performance, 34*, 41-55.

Karlsen, P. J., Allen, R. J., Baddeley, A. D., and Hitch, G. J. (2010). Binding across space and time in visual working memory. *Memory and Cognition, 38*, 292-303.

Kilb, A. and Naveh-Benjamin, M. (2007) Paying attention to binding: Further studies assessing the role of reduced attentional resources in the associative deficit of older adults. *Memory and Cognition, 35(5)*, 1162-1174.

Logie, R.H. (1995). *Visuo-Spatial Working Memory*. Psychology Press.

Logie, R. H., Brockmole, J. R., and Vandenbroucke, A. R. E. (2009). Bound feature combinations in visual short-term memory are fragile but influence long-term learning. *Visual Cognition, 17*, 160-179.

Luck, S. J., and Vogel, E. K. (1997). The capacity of visual working memory for features and conjunctions. *Nature, 390*, 279-281.

Mitchell, K. J., Johnson, M. K., Raye, C. L., and D'Esposito, M. (2000). fMRI evidence of age-related hippocampal dysfunction in feature binding in working memory. *Cognitive Brain Research, 10*, 197-206.

Mitchell, K. J., Johnson, M. K., Raye, C. L., Mather, M., and D'Esposito, M. (2000). Aging and reflective processes of working memory: Binding and test load deficits. *Psychology and Aging, 15*, 527-541.

Miyake, A., Friedman, N.P., Rettinger, D.A., Shah, P., and Hegarty, M. (2001). How are visuospatial working memory, executive functioning, and spatial abilities related? A latent-variable analysis. *Journal of Experimental Psychology: General, 130 (4)*, 621-640.

Morey, C.C., and Bieler, M. (2013). Visual short-term memory always requires general attention. *Psychonomic Bulletin and Review, 20*, 163-170.

Morey, C. C., and Cowan, N. (2005). When do visual and verbal memory conflict? The importance of working memory load and retrieval. *Journal of Experimental Psychology: Learning, Memory and Cognition, 31*, 703-713.

Moses, S. N., and Ryan, J. D. (2006). A comparison and evaluation of the predictions of relational and a conjunctive accounts of hippocampal function. *Hippocampus, 16*, 43–65.

Naveh-Benjamin, M. (2000). Adult age-differences in memory performance: Tests of an associative deficict hypothesis. *Journal of Experimental Psychology: Learning, Memory, and Cognition, 26*, 1170-1187.

Naveh-Benjamin, M., Brav, T. K., and Levy, O. (2007) The associative memory deficit of older adults: The role of strategy utilistation. *Psychology and Aging, 22(1)*, 202-208.

Naveh-Benjamin, M., Guez, J., Kilb, A., and Reedy, S. (2004). The associative deficit of older adults: Further support using face-name associations. *Psychology and Aging, 19*, 541-546.

Naveh-Benjamin, M., Hossain, Z., Guez, J., and Bar-On, M. (2003). Adult-age differences in memory performance: Further support for an associative deficit hypothesis. *Journal of Experimental Psychology: Learning, Memory and Cognition, 29*, 826-837.

Naveh-Benjamin, M. and Kilb, A. (2012) How the measurement of memory processes can affect memory performance: The case of remember/know judgments. *Journal of Experimental Psychology: Learning, Memory and Cognition, 38(1)*, 194-203.

Old, S. R., and Naveh-Benjamin, M. (2008a). Differential effects of age on item and associative measures of memory: A meta-analysis. *Psychology and Aging, 23*, 104-118.

Old, S. R., and Naveh-Benjamin, M. (2008b). Memory for people and their actions: Further evidence for an age-related associative deficit. *Psychology and Aging, 23*, 467-472.

Olson, I. R., Page, K., Moore, K. S., Chatterjee, A., and Verfaellie, M. (2006). Working memory for conjunctions relies on the medial temporal lobe. *Journal of Neuroscience, 26*, 4596-4601.

Parra, M. A., Abrahams, S., Fabi, K., Logie, R. H., Luzzi, S., and Della Sala, S. (2009). Short-term memory binding deficits in Alzheimer's disease. *Brain, 132*, 1057-1066.

Parra, M. A., Abrahams, S., Logie, R. H., and Della Sala, S. (2009). Age and binding within-dimension features in visual short-term memory. *Neuroscience Letters, 449*, 1-5.

Parra, M. A., Abrahams, S., Logie, R. H., and Della Sala, S. (2010). Visual short-term memory binding in Alzheimer's disease and depression. *Journal of Neurology, 257*, 1160–1169.

Parra, M. A., Abrahams, S., Logie, R. H., Méndez, L. G., Lopera, F., and Della Sala, S. (2010). Visual short-term memory binding deficits in familial Alzheimer's disease. *Brain, 133*, 2702–2713.

Parra, M.A., Cubelli, R., and Della Sala, S. (2011). Lack of color integration in visual short-term memory binding. *Memory and Cognition, 39*, 1187-1197.

Parra, M. A., Della Sala, S., Abrahams, S., Logie, R. H., Méndez, L. G., and Lopera, F. (2011). Specific deficit for color-color short-term memory binding in sporadic and familial Alzheimer's disease. *Neuropsychologia, 49*, 1943-1952.

Parra, M. A., Della Sala, S., Logie, R., and Abrahams, S. (2009). Selective impairment in visual short-term memory binding. *Cognitive Neuropsychology, 26*, 583–605.

Pertsov, Y., Dong, M.Y., Peich, M., and Husain, M. (2012). Forgetting what was where: The fragility of object-location binding. *PLoS One, 7 (10)*, 1-12.

Phillips, W.A. (1974). On the distinction between sensory storage and short-term visual memory. *Perception and Psychophysics, 16 (2)*, 283-290.

Piekema, C., Kessels, R. P. C., Mars, R. B., Petersson, K. M., and Fernández, G. (2006). The right hippocampus participates in short-term memory maintenance of object-location associations. *NeuroImage, 33*, 374-382.

Reuter-Lorenz, P.A., and Jonides, J. (2008). The executive is central to working memory: Insights from age, performance, and task variations. In A.R.A. Conway et al. (Eds.), *Variation in working memory.* New York: Oxford University Press.

Spencer, W. D., and Raz, N. (1995) Differential effects of aging on memory for content and context: A meta-analysis. *Psychology and Aging, 10(4)*, 527-539.

Ueno, T., Allen, R.J., Baddeley, A.D., Hitch, G.J., and Saito, S. (2011a). Disruption of visual feature binding in working memory. *Memory and Cognition, 39*, 12-23.

Ueno, T., Mate, J., Allen, R.J., Hitch, G.J., and Baddeley, A.D. (2011b). What goes through the gate? Exploring interference with visual feature binding. *Neuropsychologia, 49*, 1597-1604.

Vogel, E.K., Woodman, G.F., and Luck, S.J. (2001). Storage of features, conjunctions, and objects in visual working memory. *Journal of Experimental Psychology: Human Perception and Performance, 27 (1)*, 92-114.

West, M.J. (1993). Regionally specific loss of neurons in the aging human hippocampus. *Neurobiology of Aging, 14 (4)*, 287-293.

Wheeler, M. E., and Treisman, A. M. (2002). Binding in short-term visual memory. *Journal of Experimental Psychology: General, 131*, 48-64.

In: Working Memory
Editor: Helen St. Clair-Thompson

ISBN: 978-1-62618-927-0
© 2013 Nova Science Publishers, Inc.

Chapter 7

INTERACTIONS BETWEEN COGNITIVE LOAD FACTORS ON WORKING MEMORY PERFORMANCE IN LABORATORY AND FIELD STUDIES

Edith Galy[1] and Claudine Mélan[2]

[1]Centre de Recherche en Psychologie de la Connaissance,
du Langage et de l'Emotion, Aix-Marseille University, Marseille, France

[2]Cognition, Langues, Langage, Ergonomie,
Toulouse University, France

ABSTRACT

The aim of the present contribution is to test whether and to what extent cognitive load theory may account for working memory performance in the field of ergonomics. We will briefly present the theoretical and methodological principles of cognitive load theory, developed initially by Sweller (1988) in the field of educational psychology, and some of the more critical issues addressed recently. Thereafter, we report three experiments exploring the relations between the three load categories defined by cognitive load theory, by using working memory tasks. The influence of intrinsic and extrinsic cognitive load categories was explored by manipulating respectively task difficulty and time pressure.

The consequences of the less documented germane load were proposed to depend on participants' alertness that varies spontaneously across the day (Galy, Cariou, and Mélan, 2012). In order to test for additive effects between load categories, effects of intrinsic and extrinsic load were explored both separately and simultaneously, and when participants' alertness level was either high or low.

This procedure was tested in a mental arithmetic task and in a recall task performed by participants in controlled laboratory conditions and when the same arithmetic task was performed by air traffic controllers in a real-job situation. Results led to the conclusion of an additive interaction between intrinsic and extraneous cognitive load factors, in addition to modulatory effects by germane load.

INTRODUCTION

Working memory may be considered as the executive of thought and is as such a central concept of cognition. It enables processing and transitory storage of information and is thus crucial for a number of everyday activities, such as writing, reading, learning, and so forth. It is also critically involved in job activities requiring situation awareness and problem solving, for instance during supervision of a dynamic process, such as air traffic control or in other safety-related job situations. Miller (1956) stressed for the first time that working memory has a limited capacity, thereby raising the idea of putative cognitive overload of what was known as "short-term memory" and has been termed since as "working memory" (Baddeley and Hitch, 1974). More recent theoretical developments initiated in the 1980s by Sweller (1988) and co-workers (Sweller and Chandler, 1994; Sweller, Merrienboerd, and Paas, 1998), suggested that high mental workload would require the individual to allocate extra resources to entering information, and that the demand for extra resources may reduce processing efficiency and performance as a consequence of working memory overload. This approach is known as cognitive load theory.

Cognitive load has been defined as "… a multi-dimensional construct that represents the load that performing a particular task imposes on the cognitive system of a particular learner" (Paas and van Merriënboer, 1994, p. 122). Accordingly, cognitive load is the result of an interaction between task demands and individual characteristics. More particularly, according to Sweller (1988) three cognitive load categories may be distinguished. "Intrinsic cognitive load" depends on the intrinsic nature of the information to be processed and is thus fixed and inherent to a given task. This load category refers to the load induced by the material to be processed and by item interactivity (Ayres, 2006; Kalyuga, Ayres, Chandler, and Sweller, 2003; Sweller and Chandler, 1994). It was shown to increase more especially with task difficulty and with the number of items to be processed. "Extraneous mental workload" refers to the load induced by external factors to the task, defining the situation and the context of task execution (Sweller, 1994). External cognitive load factors can be very different and include for instance time pressure, and noise, but also work organization (Sweller, 1994). Finally, "germane mental workload" corresponds to the load allocated to working memory for the development of schemes and automatisms (Paas, Tuovinen, Tabbers, and van Gerven, 2003b; Sweller et al., 1998). Recently, Schnotz and Kirschner (2007) proposed that germane load relates to the conscious implementation of strategies reporting to meta-cognitive processes, induced by mental patterns (Sweller et al., 1998), or by conscious application of strategies to solve tasks more efficiently.

According to cognitive load theory, the three load categories are additive and the total load may not exceed available resources in order to perform a task correctly. However, relationships between cognitive load categories are asymmetric, with intrinsic load representing the baseline load on which one can yet act by decreasing, for example, task difficulty. Following attribution of resources to intrinsic load, remaining resources can be allocated to external and germane loads. If the context of task execution is characterized for instance by noise and high time pressure, much resources would be allocated to external load, and little resources would be available for germane load. However, germane load is essential to implementation of strategies enabling high task performance, in particular, when the task is difficult.

This theory has been tested repeatedly in the field of educational and learning psychology, more especially in regard to the involvement of intrinsic and extrinsic load factors on learning processes (Chanquoy, Tricot, and Sweller, 2007). Hence in order to demonstrate a more general scope of this approach, it has as yet to be examined in other research fields and in real-life situations given its decontextualized application in laboratory studies (De Jong, 2010; Moreno, 2010). Therefore, we proposed to investigate whether and to what extent cognitive load theory may further our understanding of the psychological processes involved in working memory performance, both in controlled laboratory conditions and in a real-job situation. In addition, we tested whether the less explored germane load category may be operationalized by alertness, a psychological process that is considered to be a relevant index of a subject's functional state (Galy et al., 2012). This idea is indeed supported by the finding of spontaneous alertness variations across the day (Galy, Cariou, and Mélan, 2008; Mélan, Galy, and Cariou, 2007), and by the demonstration of a high correlation between participants' body temperature and their alertness (Cariou, Galy, and Mélan, 2008).

Alertness has been shown to contribute to task performance, just like other non-specific psychological processes, including motivation or attention. Hence, it has received little attention, and has been explored in a restricted number of situations. More especially, chronobiological studies have tested the influence of alertness on other psychological processes and in particular on item processing. Mostly, studies have been performed in controlled laboratory conditions and only few attempts have been made to investigate this process in applied field studies. Hence, maintenance of a high level of alertness is of crucial interest to prevent incidents and accidents in particular in those job situations involving shift work (Folkard and Tucker, 2003). This issue will be addressed in the last section concerned with cognitive load in safety-related job activities.

ALERTNESS AS A MODULATORY FACTOR OF GERMANE LOAD IN WORKING MEMORY TASKS?

We first tested in controlled laboratory conditions the hypothesis of additive effects of intrinsic and extrinsic cognitive load factors and their relationships with alertness as a putative germane cognitive load factor.

In the literature, cognitive load measures include either subjective, performance and/or psychophysiological measures. Paas and collaborators (2003a; 2003b) reported that the reliability and the sensitivity to relatively small differences in cognitive load were particularly high for subjective rating methods. Paas and van Merriënboer (1993) stressed that performance measures may not reflect subtle changes in mental workload, and would be appropriate only if a task yields a sufficient rate of overt behavior. These authors proposed to combine performance measures and subjective measures to determine a subject's relative task efficiency. They proposed that mental efficiency could be calculated by the ratio (P-ME)/2 (ME corresponding to transformed Z-scores of cognitive effort and P to the number of correct responses). Accordingly, subjective measures would be the most sensitive measure, followed by performance and psychophysiological measures. In the present experiment cognitive load was measured by differential heart rate, subjective tension, self-reported cognitive effort, number of correct responses, and mental efficiency.

The effects of task difficulty and time pressure were explored separately and simultaneously in a working memory task. On each of 32 trials, a number was automatically generated on a computer screen and participants (30 students) had to recall item n-1 by using the keypad. Task difficulty was either low (2-digit number) or high (3-digit number); similarly, time pressure was either low (recall delay of 3500 msec) or high (recall delay 500 msec). Thus, by manipulating the two load factors separately and simultaneously four different task conditions were implemented. According to cognitive load theory, we expected task difficulty and time pressure to significantly affect cognitive load measures, and to interact with each other and with alertness.

The effects of alertness variations across the day were explored by asking participants to perform the task between 09:00 and 10:00 in the morning and seven days later, between 04:00 and 05:00 in the afternoon. Indeed, in general subjects' alertness is low in the morning and high in the afternoon (Fabbri, Natale, and Adan, 2007). In addition, several studies have reported time-of-day effects on discrimination and memory performance that were correlated with subjects' alertness but only in the most difficult task conditions (Galy et al., 2008; Mélan et al., 2007). Therefore, we expected to observe an interaction between alertness and task difficulty. As alertness and time pressure have been identified as factors affecting cognitive resources (Ayres, 2006; Backs and Seljos, 1994; Kalyuga et al., 2003; Sweller and Chandler, 1994), we also assumed an interaction between these factors. Alertness was measured by asking participants to complete Thayer's Activation–Deactivation Adjective Checklist (Thayer, 1978) on the beginning of each test session. For each of 20 listed adjectives they had to select one of several responses, "not at all", "don't know", "little" and "much", weighted respectively 1, 2, 3 and 4 (two adjectives were given negative weightings). The responses enabled us to calculate each participant's self-rated alertness that varies typically between 1 and 4.

Participants' mean self-rated alertness that was recorded in the morning and in the afternoon is represented in Figure 1 by the solid line (right-hand scale). The Figure indicates an increasing trend for participants' alertness across the day. This impression was confirmed by a regression analysis revealing significant higher alertness in the afternoon compared to the morning. Figure 1 also illustrates participants' performance in the memory task requiring them to enter item n-1 on a keypad. Performance was expressed by the mean number of correct responses (left-hand scale), when task difficulty and/or time pressure were high or low. The figure shows that the number of correct responses was at its maximum when both task difficulty and time pressure were low, while it decreased when either, or both factors, increased. The effects of task difficulty and time pressure appeared to be more marked in the morning (left panel) than in the afternoon (right panel). A multiple regression analyses revealed that performance was significantly affected by alertness and by an interaction between task difficulty and time pressure. Thus, performance decreased with low alertness, and simultaneous high difficulty and time pressure.

A similar multiple regression analysis was performed for the subjective cognitive load measures (mental efficiency, cognitive effort and tension) and for the psychophysiological load measure (heart rate variations between the test and rest periods). Significant effects were observed for mental efficiency (Figure 2) and heart rate variations (Table 1). Results indicated that the mental efficiency index was affected by the same load factors as reported for performance. Indeed, this measure decreased when alertness was low and when task difficulty and time pressure were simultaneously increased. Differential heart rate was also affected by alertness, as indicated by larger heart-rate variations when alertness was high. Multiple regression analyses performed separately for each time of day revealed a significant effect of task difficulty x time pressure on performance in the morning but not in the afternoon.

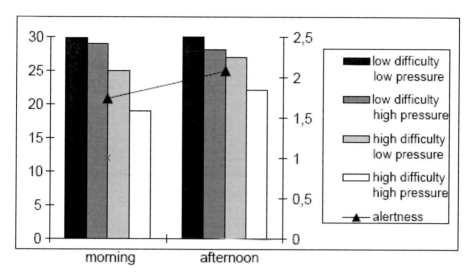

Figure 1. Number of correct responses in the morning and afternoon when participants performed a 1-back memory task, as a function of task difficulty and time pressure (left-hand scale). Solid line refers to alertness level in the morning and afternoon (right-hand scale) (Galy et al., 2012).

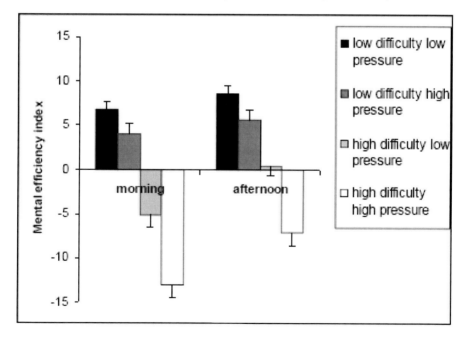

Figure 2. Mental efficiency index when participants performed a 1-back memory task, as a function of task difficulty and time pressure in the morning and afternoon (Galy et al., 2012).

Table 1. Mean heart rate variations (+/- SD) that were observed between task execution and rest-periods of equivalent duration. Data are displayed when task difficulty, time pressure and alertness were respectively low and high

		Mean heart rate variability	
Task difficulty	Low	2.29(+/- 5.14)	ns
	High	3.66(+/- 6.50)	
Time pressure	Low	1.87(+/-5.68)	ns
	High	4.08(+/-5.9)	
Alertness	Low	0.89(+/-3.96)	**
	High	5.36(+/-6.77)	

Overall, these results show that alertness interfered with subjective, behavioural and psychophysiological cognitive load measures that were considered in this experiment. In particular, when alertness was high, performance and mental efficiency increased and heart rate variations were larger. These results suggest that alertness affected the cognitive resources that were available (Fabbri et al., 2007; Galy et al., 2008; Mélan et al., 2007). Thus, with a difficult task and high time pressure, correct responses were decreased only in the morning, when cognitive resources were reduced. Conversely, task difficulty and time pressure had no effect on performance in the afternoon when cognitive resources were highest (Galy et al., 2008). Performance decrement in the morning seems to indicate that cognitive resources allocated to working memory were insufficient to execute the task efficiently when alertness was low.

These results, demonstrating differential additive effects of task difficulty and time pressure according to participants' alertness level, appear to indicate a differential contribution of germane cognitive load to task execution. When a task is easy and may be performed without any time pressure, participants would not need to use particular strategies to perform the task correctly (Schnotz and Kirschner, 2007), even when alertness is low and cognitive resources thus limited. In these conditions germane load would be virtually inexistent. On the other hand, when a task is difficult and executed under time pressure, participants must refer to adequate strategies to perform efficiently. This would then generate germane cognitive load. In other words, the limited cognitive resources in the morning would be entirely allocated to deal with the more basic intrinsic and external cognitive load factors, thereby leaving only limited resources to elaborate efficient strategies and thus for germane load. Conversely, when alertness is high, more cognitive resources may be allocated to working memory and participants can adopt useful strategies to perform a difficult task under high time pressure. Germane cognitive load is then relatively high.

In order to test this hypothesis, we explored the effects of the same load factors in a more demanding working memory task (mental arithmetic task) and by including an evaluation of participants' cognitive appraisal.

A MODULATORY EFFECT OF COGNITIVE APPRAISAL ON EXTRINSIC AND INTRINSIC LOAD FACTORS IN A WORKING MEMORY TASK?

In addition to mental load factors, mental resources available for solving a given task may be influenced by several individual characteristics, including age, morningness-eveningness, personality traits (Schnotz and Kirschner, 2007), but also situation-related aspects such as stereotype threat (Steele, 1997). Thus, an individual who belongs to a stereotyped group experiences "stereotype threat" and adopts behaviour that confirm this negative stereotype, for instance by resulting in poorer performance in a working memory task (Steele, 1997; Steele and Aronson, 1995; Steele, Spencer, and Aronson, 2002). Although numerous studies demonstrate the effects of stereotype threat on performance, questions remain about the specific mental processes that would mediate these effects. Steele and Aronson (1995) proposed that attempts to suppress stereotype-related thoughts generate anxiety and narrow subjects' attention. According to Schmader, Johns, and Forbes (2008) three interrelated factors would be involved, i.e. stress arousal, performance monitoring, and efforts to suppress negative thoughts and emotions. Evidence supporting the latter explanation was provided by the finding that stress arousal is characterized by a neurobiological activation, including increased brain activity, arterial blood pressure, cardiovascular activity, heart rate, and neuroendocrine response.

In a somewhat different approach, Folkman and Lazarus (1985) described the importance of cognitive appraisal on task performance. With a "challenge appraisal" subjects were reported to be more heavily invested in task resolution and to display better performance than with a "threat appraisal". Data from the literature suggest that task investment and cognitive resources available for additional processing appear to be determined, partly at least, by cognitive appraisal. The present experiment was designed to test this hypothesis by using an arithmetic task, known to be sensitive to cognitive stereotyping (Wraga, Helt, Jacobs, and Sullivan, 2006), and that may provide an indication of the participants' confidence in their ability to perform the task (Steele, 1997).

The 31 participants (Mean age = 23.3; SD = 1.86) that volunteered to take part in this experiment were biology (9 women and 9 men) and psychology students (6 women and 7 men). Each subject performed a mental arithmetic task in four experimental conditions that varied in respect to task difficulty and time pressure, like in the first experiment. Task difficulty was manipulated by requiring participants to add either 5 (low task difficulty) or 36 (high task difficulty) to a number displayed on the screen. Two- and three-digit numbers were selected so that in each condition the algebraic sum changed the tenths or the hundredths on half of the 32 trials. In the low time pressure conditions, there was no time limit to enter the sum on a numeric keypad, even though subjects were instructed to respond as quickly and accurately as possible. In the high time pressure conditions, the time limit was set at 8,000 ms. The same cognitive load measures as in the first experiment were recorded (differential heart rate, mental effort, subjective tension, cognitive efficiency index, subjective alertness, number of correct responses). To measure participants' cognitive evaluation of the task they rated the perceived task-induced stress and their resources to perform the task, respectively on two visual analogue scales. The ratio of the two measures indicated their cognitive appraisal, as described by Tomaka and co-workers (1993).

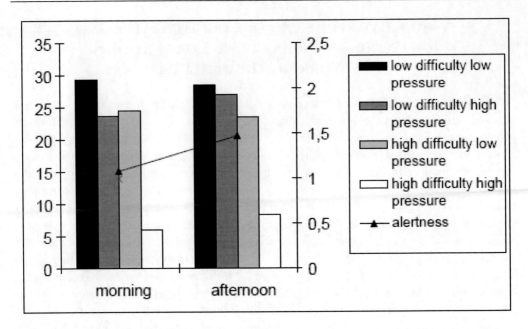

Figure 3. Mean number of correct responses (left-hand scale) when participants performed a mental arithmetic task in each of four experimental conditions, both in the morning (left-hand panel) and in the afternoon (right-hand panel). The solid line indicates participants' mean self-rated alertness measured in the morning and in the afternoon (right-hand scale).

Results showed that alertness, represented on Figure 3 by the solid line, was lower in the morning than in the afternoon, and affected performance as the number of correct responses increased with alertness. While alertness did not affect either subjective measure, it interfered with differential heart rate as indicated by greater heart rate variations between the test and rest periods when participants' alertness was high.

A significant interaction between task difficulty and time pressure indicated that performance decreased when task difficulty and time pressure were simultaneously increased (Figure 3). A similar interaction between task difficulty and time pressure affected perceived cognitive effort, and mental efficiency, summarized in Table 2. These results indicate that the greater task difficulty and time pressure, the higher subjective load measures, and the lower mental efficiency.

A task difficulty x time pressure interaction affected the number of correct responses in the morning but not in the afternoon, and in subjects with a "threat appraisal" but not in those with a "challenge appraisal". When these two factors (cognitive appraisal and time of day) were included in the same analysis, a significant interaction between task difficulty and time pressure reduced the number of correct responses for both "threat appraisal" subjects and "challenge appraisal" subjects in the morning, but only for "threat appraisal" subjects in the afternoon.

Table 2. Subjective load measures (Mean +/- SD) rated by participants following the mental arithmetic task as a function of task difficulty and time pressure, both in the morning and in the afternoon. Upper panel shows cognitive effort. Lower panel shows mental efficiency

	Mean cognitive effort ratings			
	low difficulty		high difficulty	
	low pressure	high pressure	low pressure	high pressure
morning	5.18 (+/-2.73)	6.56 (+/-2.43)	7.91 (+/-1.07)	8.78 (+/-1.09)
afternoon	3.22 (+/-2.44)	4.36 (+/-2.68)	7.37 (+/-1.98)	8.55 (+/-1.16)

	Mean mental efficiency index			
	low difficulty		high difficulty	
	low pressure	high pressure	low pressure	high pressure
morning	7.23 (+/-5.71)	1.18 (+/-5.63)	-1.02 (+/-3.73)	-13.61 (+/-2.85)
afternoon	10.46 (+/-4.85)	7.30 (+/-6.39)	-0.61 (+/-5.36)	-11.31 (+/-4.63)

The present results differed from those reported above concerning the impact of load factors on perceived cognitive effort ratings. In the present experiment, self-rated cognitive effort was modulated by a task difficulty x time pressure interaction, while a main effect of task difficulty occurred in the previous study. In other words, participants would have perceived time pressure as a greater constraint when calculating the sum of two numbers than when retrieving numbers from short-term memory. This may be explained by the fact that the arithmetic task was cognitively more demanding than the memory task, given that in the former case subjects had to manipulate the numbers in addition to maintaining them in the short-term buffer, whereas in the latter case they only had to maintain the numbers. It thus seems reasonable to conclude that participants gained more readily the impression of reaching their cognitive limits in the former case.

Performance was also affected by an interaction between task difficulty and time pressure in subjects that had a "threat appraisal" approach, but not in their "challenge appraisal" counterparts.

A similar pattern of results was obtained for alertness variations, as performance decreased with increasing intrinsic and extrinsic factors only when participants' alertness was low. It may be argued that the low task difficulty and time pressure condition did probably not require any particular strategies for performing the task, while this may have been necessary in the high difficulty and time pressure condition. If this was indeed the case, then it is plausible to find that factors known to affect performance (cognitive appraisal and alertness) decrease performance only in the more demanding task conditions (Galy et al., 2008; Mélan et al., 2007).

According to Folkman and Lazarus (1985), cognitive appraisal determines task investment and the use of available cognitive resources for additional strategic cognitive processing. Thus, subjects with a "threat appraisal" are thought to display lower task investment than those with a "challenge appraisal". Accordingly, the latter would have been able to use more appropriate strategies, as a result of their task investment in particular in the demanding task condition.

Taken together, the present findings lend further support to the model represented in Figure 4 that will be discussed below. However, beforehand we will test whether the relations between cognitive load factors observed in the two experiments performed in controlled laboratory conditions may be generalized to real-job conditions.

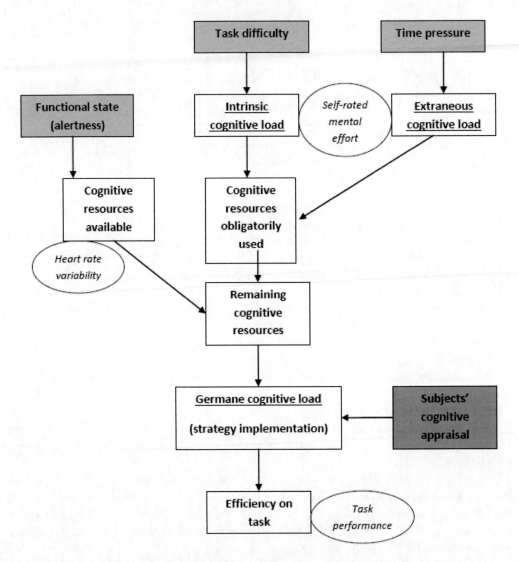

Figure 4. Graphical representation of putative relationships between cognitive load factors and cognitive load categories in working memory tasks (Galy et al., 2012).

COGNITIVE LOAD FACTORS, ALERTNESS AND JOB PERCEPTION

A number of controlled laboratory studies and some field studies have explored the way task factors such as presentation modality, number of items, and type of information impact on memory performance. However, in real-job situations these factors interact with task-independent factors, and in particular with work schedule features. Air Traffic Controllers (ATCs) mnemonic performance has been shown to be affected both by task-dependent factors (item modality and number of items), and by task-ipendent factors (alertness and the number of hours on-shift) (Galy, Mélan, and Cariou, 2006; Galy et al., 2008; 2010; Mélan et al., 2007). Like-wise, some variables, more dependent on external stress factors, in particular perceived tension and heart rate, have been shown to vary both with time-of-day and with work load (Cariou et al., 2008). In light of these findings we explored the effects on working memory performance of two factors that are largely involved in ATC control activities (i.e. task difficulty and time pressure) and the way performance is related to subjective variables (self-rated alertness and tension), and to perceived job characteristics. More specifically, time pressure has proved to be one of the most common stressors in the work environment, where time may be part of a mediating process that influences perception of control (Koslowsky, Kluger, and Reich, 1995). Thus, increased heart rate is commonly reported in situations involving high tension and/or mental effort in real or simulated work conditions (Boucsein and Ottmann, 1996; Ritvanen, Louhevaara, Helin, Väisänen, and Hänninen, 2006), and has been shown to vary as a function of self-reported stress (Sloan, Shapiro, Bagiella, and Boni et al., 1994; Ritvanen et al., 2006). In consequence, we tested whether cognitive load theory could account for results obtained in a real-life situation, given that some authors have stressed its decontextualized application in laboratory studies (De Jong, 2010; Moreno, 2010).

The impact of psychosocial stress at work could be an important additional or mediating factor for performance, and, in the long run, for health. The most important models for job stress research are "the job demand-control model" (Karasek, 1979) and "the effort-reward imbalance model" (Siegrist, 1996). The job demand-control model or job strain model operationalizes two dimensions of stressful work: Job demands, defined as psychological stressors present in the work environment (e.g. high pressure resulting from time lines), and job control, referring to workers' **opportunities** to make use of their own skills and abilities, to learn something new at work and to control their tasks, as opposed to monotonously acting on orders in social isolation (Karasek, Brisson, Kawakami, Houtman, Bongers, and Amick; 1998; Vogel, Braungardt, Meyer, and Schneider, 2012). Social support at work is a further dimension of this model, assuming that the most risky types of jobs are characterized by high quantitative demands combined with low control and low support at work. The authors suggest that such adverse working conditions could lead to the development of psychological and physical illness. Accordingly, we showed previously that psychosocial job characteristics varied along with sleep and health in air traffic controllers and satellite controllers (Mélan, Cariou, Galy, and Cascino, 2008).

Air traffic controllers participating in the present experiment (9), aged between 34 and 56 years old, worked on day-shifts starting either at 08:30 or 10:00 in a flight test control centre in the South-West of France. They were trained by the French Air Force to the specific control task and had a ten-year experience at least. Briefly, they were seated in front of typical flight control equipment (radar screen, strips) and had to determine a temporal and spatial gap enabling to test functional properties of commercial aircrafts before delivery to the customer, as well as of prototypes. A test trial lasted between 20min and 12h, but mostly one hour. The timetable organizing control activities of the day were in principle provided the day before, but required frequent adjustments.

Controllers were submitted to the experimental procedure described in the previous section on three occasions of the shift (1h after shift-beginning, in the middle of the shift and 1hr prior shift-end). Briefly, on each test session, they first rated Thayer's (1978) adjective checklist enabling to calculate their alertness and tension indexes, then performed the mental arithmetic task exploring the effects of intrinsic (task difficulty) and extrinsic (time pressure) cognitive load factors according to the four task conditions described above. Task difficulty was manipulated by asking participants to add either 5 (low task difficulty) or 36 (high task difficulty) to a two- and three-digit number displayed on a screen. Similarly, time pressure was either low (no time limit to enter the sum on a numeric keypad, even though subjects were instructed to respond as quickly and accurately as possible) or high (8,000ms for responding). Task-performance was expressed by participants' response latency (delay before to entering the sum on the keypad), and by the amount of non-responses as a consequence of the limited response delay. Finally, participants completed a questionnaire comprised of 56 questions exploring several job characteristics, and in particular job demands, job control and social support at work, by using items related to those contained in the Job Content Questionnaire (Cascino et al., submitted; Karasek et al., 1998). Results are shown in Figures 5, 6 and 7.

As indicated in Figure 5, when participants performed the mental arithmetic task, their response latencies and number of non-responses varied significantly according to task factors ($F_{1,24}=277,13$; $p<10-3$; $F_{1.24}=64.41$; $p<10-3$). Both measures increased when time pressure and task difficulty increased, either separately or simultaneously. Thus, performance decreased significantly when either task difficulty or time pressure increased compared to the simple task condition, and also when both factors increased simultaneously rather than either of them (($p<10-3$ in each case).

Investigation of ATCs' self-rated alertness and tension at the different times of the day (Figure 6) revealed that their alertness decreased slightly, but non-significantly across the shift, while their perceived tension remained stable throughout the shift. These subjective measures were however related to controllers' performance, as indicated by a positive correlation of alertness and tension with response latency, respectively when task difficulty was high alone or together with time pressure.

Taken together, these results indicate additive effects of time pressure and task difficulty on ATCs' on-job performance, thereby extending to a real-job situation the data obtained in controlled laboratory conditions. Further, alertness and tension were correlated with task performance. Hence, subjective alertness, an indicator of subjects' functional state (Cariou et al., 2008), has been shown to contribute to the incident and accident rates in real-job situations (Folkard and Tucker, 2003). Therefore, the results of this field study have been interpreted in light of current models of cognitive load factors (Sweller, 1998), attributing a modulatory effect of participants' alertness on germane cognitive load (Galy et al., 2012). It is noteworthy that ATCs perceived their working conditions as an active job situation, with low job-strain. In addition, they perceived little work-family conflicts and high availability of technical resources at work, thereby raising the possibility that these additional dimensions may have contributed to ATCs' job perception. However, further research is necessary to highlight the way psychosocial stress at work may influence job performance.

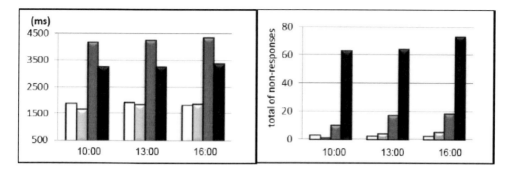

Figure 5. ATCs' performance in a mental arithmetic task with low task difficulty and time pressure (white bars), low task difficulty and high time pressure (light grey bars), high task difficulty and low time pressure (dark grey bars), and both high task difficulty and time pressure (black bars). Left panel: Mean response latencies to enter the sum of two numbers on a keypad in each task condition. Right panel: Total of non-responses on the 32 trials in each task condition.

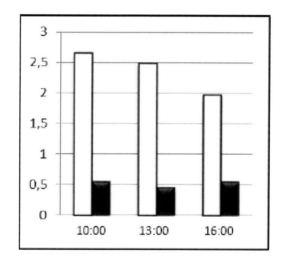

Figure 6. Mean self-rated alertness (white bars) and tension (black bars) on three occasions of the shift.

Figure 7 shows that ATCs rated a high level of control and of psychological demands, associated with high social support by their colleagues. The combination of high demands, control and support according to Karasek's model characterizes an active type of job with low job-strain. Further, job-family conflicts were almost absent, while available resources were perceived as being very high; both characteristics may thus have contributed to define a satisfactory job situation.

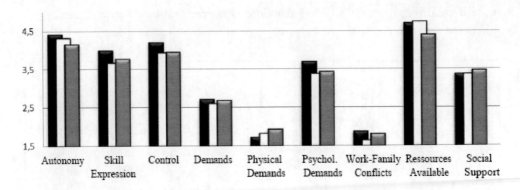

Figure 7. Psychosocial job characteristics rated by ATCs on a 56-item questionnaire (Cascino et al., submitted), 1h after shift beginning (black bars), on the middle of the shift (white bars) and 1h before shift-end (grey bars).

CONCLUSION

Taken together, results of the present laboratory and field studies confirm the additive (Pass et al., 2003b) but asymmetric nature of the relationship between germane load on one hand, and intrinsic and extraneous loads on the other (Schnotz and Kirschner, 2007). Germane load is determined by the subject's choice of strategies to be applied to the task (as a function of alertness and cognitive appraisal). The results stress the recent view of de Jong (p.113, 2010) "one might say that intrinsic and extraneous cognitive load concern cognitive activities that must unavoidably be performed, so they fall under cognitive load; germane cognitive load is the space that is left over that the learner can decide how to use, so this can be labelled as cognitive effort". They further suggest that in the field of cognitive psychology, implementation of subjects' strategies is determined by the amount of free cognitive resources which, in turn, is determined by intrinsic load (task difficulty) and extraneous load (time pressure).

To conclude, the present studies stress the importance of addressing cognitive load theory and working memory requirements in the field of ergonomics. The consideration of three mental workload categories was necessary to understand differences on measures of mental workload changes (Figure 4). The effects of task difficulty and time pressure were shown to be additive in working memory tasks, which may explain why these factors are often cited in relation to wellbeing and safety issues at work. However, their combined action on mental workload measures was modulated by alertness, suggesting that it may be useful to consider the overall situation, rather than focusing only on specific task characteristics. Thus, in a real-job situation, mental overload can be the result of a combination of task characteristics, such as time pressure and task difficulty, but its occurrence appears to depend on job characteristics, in particular shift scheduling and the job-situation (Mélan et al., 2008; Siegrist, 2010). As a consequence, solutions designed to reduce incidents and accidents at work must take organizational considerations into account as well as more technical aspects.

REFERENCES

Ayres, P. (2006). Using subjective measures to detect variations of intrinsic cognitive load within problems. *Learning and Instruction* 16, 389–400.

Backs, R.W., and Seljos, K.A. (1994). Metabolic and cardio respiratory measures of mental effort: the effects of level of difficulty in a working memory task. *International Journal of Psychophysiology* 16 (1), 57–68.

Baddeley, A.D., and Hitch, G.J. (1974). Working memory. In G.H. Bower (Ed.), *The psychology of learning and motivation (Vol 8, pp.* 47-89). London: Academic Press.

Boucsein,W., and Ottmann, W. (1996). Psychophysiological stress effects from the combination of night-shift work and noise. *Biological Psychology* 42 (3), 301–322.

Cariou, M., Galy, E., and Mélan, C. (2008). Differential 24-h variations of alertness and subjective tension in process controllers: investigation of a relationship with body temperature and heart rate. *Chronobiology International* 25(4), 597-609.

CASCINO, N., GALY, E., MÉLAN, C. (submitted) Perception of job characteristics across the shift and according to the shift worked. European Revue of applied Psychology.

Chanquoy, L., Tricot, A., and Sweller, J. (2007). La charge cognitive. Paris: A. Colin.

De Jong, T. (2010). Cognitive load theory, educational research and instructional design : some food for thought. *Instructional science*, 38, 105-134.

Fabbri, M., Natale, V., and Adan, A. (2007). Effect of time of day on arithmetic fact retrieval in a number-matching task. Acta Psychologica 127 (2), 285–290.

Folkard, S., and Tucker, P. (2003). Shiftwork, safety and productivity. *Occupational Medecine* 53, 95-101.

Folkman, S., and Lazarus, R. S. (1985). If it changes it must be a process: Study of emotion and coping during three stages of a college examination. *Journal of Personality and Social Psychology* 48(1), 150-170.

Galy E, Cariou M., Mélan C. (2012). What is the relationship between mental workload factors and cognitive load types? *International Journal of Psychophysiology* 83, 269–275.

Galy, E., Mélan, C., and Cariou, M. (2008). Investigation of task performance variations according to task requirements and alertness across the 24-h day in shift-workers. *Ergonomics* 51(9),1338-51.

Galy, E., Mélan, C., and Cariou, M. (2010). Investigation of ATCs' response strategies in a free recall task: what makes auditory recall superior to visual recall? *International Journal of Aviation Psychology.* 20(3), 295-307.

Kalyuga, S., Ayres, P., Chandler, P., and Sweller, J. (2003). The expertise reversal effect. *Educational Psychologist* 38, 23–31.

Karasek, R. (1979). Job demands, job decision latitude, and mental strain: Implications for job redesign. *Administrative Sciences Quaterly* 24, 285-308.

Karasek, R., Brisson, C., Kawakami, N., Houtman, I. Bongers, P., and Amick, B. (1998). The Job Content Questionnaire (JCQ): An instrument for internationally comparative assessments of psychosocial job characteristics. *Journal of Occupational Health Psychology,* 3 (4), 322-355.

Koslowsky, M., Kluger, A.N., and Reich, M. (1995). Commuting Stress: Causes, Effects, and Methods of Coping. Plenum, New York.

Mélan, C., Cariou, M., Galy, E., and Cascino, N. (2008). Shift-work, psychosocial job characteristics, sleep and health in air traffic controllers and satellite controllers. Eu Assoc Aviat Psychol. http://www.eaap.net/fileadmin/Eaap/downloads/28EAA PValencia 2008/PROCEEDINGS_28th_EAAP_-_standard.pdf (pp. 205-210).

Mélan, C., Galy, E., and Cariou, M. (2007). Mnemonic Processing in Air Traffic Controllers (ATCs): Effects of Task Parameters and Work Organization. *International Journal of Aviation Psychology* 17(4), 391-409.

Miller, G.A. (1956). The magical number seven, plus or minus two: Some limits on our capacity for processing information. *Psychological Review* 63, 81-97.

Moreno, R. (2010). Cognitive load theory: more food for thought. *Instructional science* 38, 135-141.

Paas, F.G.W.C., Renkl, A., and Sweller, J. (2003a). Cognitive load theory and instructional design: recent developments. *Educational Psychologist* 38 (1), 1–4.

Paas, F.G.W.C., Tuovinen, J.E., Tabbers, H., and van Gerven, P.W.M. (2003b). Cognitive load measurement as a means to advance cognitive load theory. *Educational Psychologist* 38, 63–71.

Paas, F.G.W.C., and van Merriënboer, J.J.G. (1993). The efficiency of instructional conditions: an approach to combine mental-effort and performance measures. *Human Factors* 35, 737–743.

Paas, F.G.W.C., and van Merriënboer, J.J.G. (1994). Variability of worked examples and transfer of geometrical problem solving skills: a cognitive load approach. *Journal of Educational Psychology* 86, 122–133.

Ritvanen, T., Louhevaara, V., Helin, P., Väisänen, S., and Hänninen, O. (2006). Responses of the autonomic nervous system during periods of perceived high and low work stress in younger and older female teachers. *Applied Ergonomics* 37, 311–318.

Schmader, T., Johns, M., and Forbes, C. (2008). an integrated process model of stereotype threat effects on performance. *Psychological Review* 115(2), 336-356.

Schnotz, W., and Kirschner, C. (2007). A reconsideration of cognitive load theory. *Educational Psychology Review* 19, 469–508.

Siegrist J. (1996). Adverse Health Effects of High-Effort/Low-Reward Conditions. *Journal of Occupational Health Psychology* 1, 27-41.

Siegrist J. (2010). Effort-reward imbalance at work and cardiovascular diseases. *International Journal of Occupational Medicine and Environmental Health* 23, 279-285.

Sloan, R.P., Shapiro, P.A., Bagiella, E., Boni, S.M., Paik, M., Bigger, J.T., and Gorman, J.M. (1994). Effect of mental stress throughout the day on cardiac autonomic control. *Biological Psychology* 37, 89–99.

Steele, C. M. (1997). A threat in the air. How stereotypes shape intellectual identity and performance. *American Psychologist* 52, 613-629.

Steele, C. M., and Aronson, J. (1995). Stereotype threat and the intellectual test performance of African Americans. *Journal of Personality and Social Psychology* 69, 797-811.

Steele, C. M., Spencer, S. J., and Aronson, J. (2002). Contending with group image: The psychology of stereotype and social identity threat. In M. Zanna (Ed.), *Advances in experimental social psychology* (Vol. 34, pp. 379-440). San Diego: Academic Press.

Sweller, J. (1994). Cognitive load theory, learning difficulty and instructional design. *Learning and Instruction* 4, 295-312.

Sweller, J. (1988). Cognitive load during problem solving: effects on learning. *Cognitive Science* 12, 257-285.

Sweller, J., and Chandler, P. (1994). Why some material is difficult to learn. *Cognition and Instruction* 12, 185-233.

Sweller, J., Van Merriënboer, J.J.G., and Paas, F.G.W.C. (1998). Cognitive architecture and instructional design. *Educational Psychology Review* 10, 251-296.

Thayer, R.E.. (1978). Toward a psychological theory of multidimensional activation (arousal). *Motivation and Emotion* 2, 1–34.

Tomaka, J., Blascovich, J., Kelsey, R. M., and Leitten, C. L. (1993). Subjective, physiological, and behavioral effects of threat and challenge appraisal. *Journal of Personality and Social Psychology* 65, 248-260.

Vogel, M., Braungardt, T., Meyer, W., and Schneider, W. (2012). The effects of shift work on physical and mental health. *Journal of neural transmission* 119(10), 1121-32.

Wraga, M., Helt, M., Jacobs, E., and Sullivan, K. (2006). Neural basis of stereotype-induced shifts in women's mental rotation performance. *Social Cognitive and Affective Neuroscience* 2(1), 12-19.

In: Working Memory
Editor: Helen St. Clair-Thompson

ISBN: 978-1-62618-927-0
© 2013 Nova Science Publishers, Inc.

Chapter 8

AFFECTIVE WORDS INFLUENCE PROCESSING IN VISUAL AND AUDITORY WORKING MEMORY

Fumiko Gotoh
Bunkyo University, Saitama Prefecture, Japan

ABSTRACT

A variety of definitions of working memory (WM) have been published that all converge in conceptualizing WM as an active mental manipulation system with a limited capacity for temporarily maintaining information. Although the relationship between cognition and emotion has received significant research attention recently, WM models are still premised on the notion of processing affectively neutral information. Therefore, how WM processes emotion-laden information, such as, for instance, language is unclear. However, in everyday life, we are often exposed to emotion-laden information. If WM is indeed a central processing system, it is important to reveal how that system processes emotional (also referred to as *affective*) information. On this background, the aim of this chapter is to contribute to the understanding of how the affective valence of written or spoken words influences processing in WM. The processing of words involves visual or auditory perception for retrieving semantic information and long-term memory (LTM) for accessing valence and connotation-related knowledge, which fits well with Cowan's WM model (1999). The research reported here (Gotoh, 2008, 2012; Gotoh, Kikuchi, and Roßnagel, 2008; Gotoh, Kikuchi, and Olofsson, 2010) revealed that WM is automatically influenced by the affective valence of words (negative and positive) in various emotional WM tasks in both visual and auditory WM. Consequently, emotion-laden written and spoken words induced dominance, interference, and facilitation effects compared to neutral words, in that response times or accuracy are diminished and facilitated, respectively. In addition, we revealed a factor that possibly distinguishes between affective interference and facilitation effects, inducing both effects within the same task (Gotoh et al., 2010). Importantly, the affective influence may be due to a central attention mechanism, rather than to an attentional anxiety bias or simply due to practice. Finally, the influence of affective valence of words emerged for both the visual and the auditory processing of WM. Thus, the central processing system does not invariantly process all incoming stimuli but it may be affectively modulated, regardless of the visual or auditory nature of stimuli.

INTRODUCTION

A variety of definitions of working memory (WM) have been published (Miyake and Shah, 1999). For instance, Baddeley conceptualized WM as multiple specialized components of cognition (e.g., Baddeley and Logie, 1999) and Cowan conceptualized WM as embedded processes (Cowan, 1999). Although the focusing mechanism varies between researchers, a clear consensus with WM is of an active mental manipulation system with a limited capacity for temporarily maintaining information. In terms of this function, WM differs from earlier concepts of short-term memory in a significant way.

In WM systems, a critical and controversial component is the central executive that is assumed to regulate or control the system and execute the processing functions of WM. According to Smith and Jonides (1999), even though there is a lack of consensus about the central executive's function, some agreements include; 1) attention and inhibition: focusing attention on relevant information and processes and inhibiting irrelevant ones, 2) task management: scheduling processes in complex tasks, which requires the switching of focused attention between tasks, 3) planning: planning a sequence of subtasks to accomplish some goal, 4) monitoring: updating and checking the contents of WM to determine the next step in a sequential task, and 5) coding: coding representations in WM for time and place of appearance.

Many imaging studies have revealed that the prefrontal cortex area corresponds to WM (Braver et al., 1997; Smith, Jonides, and Koeppe, 1996). Specially, the dorsolateral prefrontal cortex (DLPFC) is thought to correspond to the central executive (Cohen et al., 1997). In addition to the DLPFC, the anterior cingulate cortex (ACC) is also thought to correspond to the central executive (Smith and Jonides, 1999; Todd and Marois, 2004).

WM is also generally considered to be of limited capacity and to contribute to cognition by activating and maintaining task-relevant information while inhibiting task-irrelevant information (e.g., Cowan, 2005; Miyake and Shah, 1999). Accordingly, many studies have revealed a significant relationship between WM and cognitive abilities, for instance, reading ability (Daneman and Carpenter, 1980), language comprehension (King and Just, 1991; MacDonald, Just, and Carpenter, 1992), reasoning (Kyllonen and Christal, 1990), and intelligence (Conway, Cowan, Bunting, Therriault, and Minkoff, 2002; Daneman and Merikle, 1996; Engle, Tuholski, Laughlin, and Conway, 1999). With respect to capacity, Miller proposed that it would be seven items (Miller, 1956). Recently, however, Cowan (Cowan, 2001) proposed that it would be four items.

Although the relationship between cognition and emotion has received significant research attention recently, for instance, on research on memory (Berthoz, Blair, Le Clec'h, and Martinot, 2002; Colombel, 2000; Christianson, 1992; Doerksen and Shimamura, 2001; Hamann, 2001; Reisberg and Hertel, 2004; Reisberg and Hertel, 2004) and attention (e.g., Fox, Russo, Bowles, and Dutton, 2001; Williams, Mathews, and MacLeod, 1996; Yiend and Mathews, 2001), WM models are still premised on the notion of processing affectively neutral information. Therefore, how WM processes emotion-laden information, such as, for instance, language is unclear. However, in everyday life, we are often exposed to emotion-laden information. If WM is indeed a central processing system, it is important to reveal how that system processes emotional (also referred to as affective) information.

For answering that question, I explored how the system processes the affective valence of language, comparing affective words of negative, positive, and neutral connotation. Processing language requires higher-order cognition, for instance, matching written or spoken words with the lexicon in long-term memory (LTM) in order to recognize words and process their connotation. WM models represent our mental work space including the influence of LTM, such as affective valence. Cowan's model (e.g., Cowan, 1999) conceptualizes WM as the temporarily activated portion of LTM. Consequently, when emotion-laden language is encoded in WM, its connotation is elicited by the affective valence stored in LTM. To obtain affective words, we assessed in a survey on a 7-point scale the valence of a series of words, defining as negative and positive valences, respectively, significant differences from 4 as the neutral scale midpoint (Gotoh and Ohta, 2001). The affective words used for my experiments were two-compound kanji words, one-compound kanji words, or auditory words. The words were also surveyed for word frequency, imagery, and arousal.

Using affective words, crucial questions concern the invariance of affect effects and how WM processes affective information. If WM is purpose-built to, for instance, reach a goal or solve a task, affective valence of information may be treated the same as neutral information.

INFLUENCE OF AFFECTIVE VALENCE ON VISUAL WORKING MEMORY

As mentioned above, my research attempted to reveal the impact of affective valence on WM. The first attempt was to reveal the influence of affective valence on visual WM (Gotoh, 2008; Gotoh, et al., 2008; Gotoh et al., 2010). The research revealed an affective interference effect using three WM tasks with written affective words. Response times or accuracy were diminished after processing affective words compared to neutral words. In contrast, affective facilitation also emerged with the affective interference effect within the same task (Gotoh et al., 2010), suggesting possible factors that distinguish affective interference and facilitation effects.

Affective Interference Effects on Visual WM: Evidence from a Digit-Updating Task (Gotoh, 2008)

As mentioned, WM is a limited-capacity system for holding and manipulating information for relatively short time periods (e.g., seconds). Cowan (1999) conceptualized WM as a temporarily activated portion of LTM, arguing that capacity limits of WM arise because only four or so items within this activated portion of LTM can be in the focus of attention at any given time. Extending this model, Oberauer (2002) conceptualized WM as a concentric set of regions in LTM with three distinct stages: the activated part of LTM, the region of direct access holding a limited number of items available for cognitive processing, and the focus of attention holding only one item actually selected for processing. Oberauer (2002) referred to time costs for shifting the focus of attention from one mental object to another within the active set as object switching costs. Since in Oberauer's framework the focus of attention holds only one item for processing, object switching costs were found when that item had to be switched. In this view, affective valence of material held in WM is likely to have its greatest influence on WM when the material has to be switched.

Previous research has revealed that threat-related words (e.g., death, disease, failure) are more likely to capture attention than emotionally neutral words for both anxious and nonanxious individuals (McNally, Riemann, and Kim, 1990; Segal, Gemar, Truchon, Guirguis, and Horowitz, 1995). In addition, it has been shown that threat-related stimuli make it difficult to disengage attention from the threatening stimulus, and that this effect interacts with anxiety level (Fox et al., 2001; Yiend and Mathews, 2001).

Taking this evidence and Oberauer's (2002) model together, it appears that WM tasks will be processed more slowly if individuals are required to switch attention from threat-related or negative items to neutral items during the course of the WM task. To test this proposition, Gotoh (2008) asked participants to perform a task in which word recognition and WM updating (Oberauer, 2002; Salthouse, Babcock, and Shaw, 1991) were required alternately. The study used two-compound affective words (negative and positive words) and neutral words for the recognition task. On each task, participants first memorized two digits and two kanji words. Next, they performed to switch between word recognition and WM updating tasks. The idea was that if affective words capture attention, it would be reflected in the updating task, compared with neutral words. In addition, an anxiety inventory was administered to assess the influence of anxiety on the emotional interference effect (Shimizu and Imae, 1981; Spielberger, Gorsuch, and Lushene, 1970).

Results demonstrated that affective valence influences the attention switching or attention disengagement component (see Fox et al., 2001; Oberauer, 2002) of WM. Across two experiments, the results showed that performance on the WM updating task was slower following negative versus neutral recognition words. In contrast, positive words did not show a constant influence. Accordingly, the results demonstrated the affective interference effect for negative valence. In addition, the results revealed that the affective interference effect was not modified by level of anxiety. Therefore, the results allow for concluding that the switch cost for affective valence is not due to the attention bias of anxiety but it could be due to the central attention mechanism.

Affective Interference Effects on Visual WM: Evidence from a Counting Task (Gotoh Et Al., 2008)

In a further experiment, Gotoh et al. (2008) attempted to confirm the affective interference effect on WM in another task. We used affective one-compound kanji words for a counting task. The task was to subitize or count the words presented on the computer screen. On each trial, a set of 1 to 10 affectively negative, neutral, or positive words were presented for 200 ms each. The presented words were always of the same valence on each trial. Participants counted the words after each trial.

Before the results, I explain the two processes subitizing and counting. When people are presented with a number of items to enumerate they engage one of two processes, depending on the number of to-be-enumerated visual objects. In subitizing (Kaufman, Lord, Reese, and Volkmann, 1949), the number of objects can be "seen at a glance" without iterative processes and then, this process results in fast, effort-free, and accurate enumeration of less than four items. In counting, in contrast, the number of objects is serially adjusted in one's internal count upon perceiving an item and then, the process results in relatively slow, effortful, and error-prone enumeration of more than four items.

According to Mandler and Shebo's (1982) results, there appears to be a countable region (about 1 to 6 or 7 items) involving the subitizing and counting regions, and an uncountable estimation region (more than 8 items). By contrast, Cowan (2005) argued that enumerating data can be explained by three regions: a subitizing region (about 1 to 4 items), a counting region (about 4 to 8 items), and an estimation region (more than eight or so items). In terms of affective stimulus processing, the countable region is likely to be crucial. In the countable region, each stimulus might be recognized and word connotation might be processed, enabling one to give the correct answer. In the uncountable region, however, every single stimulus might not be recognized, limiting the influence of connotation on the central executive.

Based on these considerations, we analyzed the enumerating data separately by the countable region (1 to 7 items), including the subitizing range (1 to 4 items) and the counting range (5 to 7 items), and the uncountable region (8 to 10 items), as a function of affective valence. We found an affective interference effect, expressed in longer response times for negative and positive words relative to neutral words within the subitizing and counting rages. In addition, we obtained increased error rates within the counting range. In line with our expectations, affective interference was restricted to set sizes within the countable range. The interference effect did not differ between positive and negative words.

Thus, the results replicated the affective interference effect on WM using affective words. Unlike recognition of the words, the counting task is not explicitly required to process each word but it required to enumerate the number of the presented words. Even under this condition, word connotation influences WM processing. Therefore, it is thought to be an automatic influence of affective valence on WM processing. In addition, since the affective interference effect was shown in the countable region, it seems that the capacity of WM relates to the affective interference effect.

Conclusions for Affective Interference Effects on Visual WM

The results in the two tasks demonstrated the effect of valence on WM. Negative words induced the affective interference effect on WM processing, showing slower response time and/or increased error. The result demonstrated that the affective interference effect was not due to an attention bias of anxiety. Therefore, the effect could be from the central executive, in which valence influences the attention switching or attention disengagement component (Fox et al., 2001; Oberauer, 2002) of WM. Following this, in the updating task if negative valence did not influence attention switching or an attention disengagement component, response time would not differ between negative and neutral words. Likewise, in the counting task if affective valence did not influence the attention component, enumeration would not differ between affective (negative and positive) and neutral words.

The affective interference effect emerged not only in the recognition task but also in the counting task. This indicates that valence may automatically influence the attention mechanism of WM. In the recognition task, the meaning of the words may be used for the task. Therefore, affective valence plays some role in the recognition task. However, in the counting task, the task did not require recognizing the meaning of the presented words. Even in this situation, the affective valence influenced the attention mechanism and induced the affective interference effect. Therefore, it could mean that affective valence automatically influences the attention mechanism of WM.

Regarding negative and positive valence, negative valence always induced the affective interference effect in the two tasks. Positive valence, however, did not show a consistent influence. This could be explained by a *negativity bias* that is assuming that negative, but not positive stimuli interrupt processing due to their relevance for survival (e.g., Pratto and John, 1991). It seems that negative valence therefore interferes with ongoing central mechanisms in WM due to their relevance for survival.

The Affective Facilitation Effect on Visual WM: Evidence from a Retro-Cue Memory Task (Gotoh Et Al., 2010)

In contrast with the previous two studies, in a retro-cue memory task (Gotoh et al., 2010), we examined the affective facilitation effect as well as the affective interference effect on WM tasks. We expected to show both facilitation and interference within one task and reveal a factor that distinguishes between them. Although valence was shown to have an interference effect on WM, the rationale that we explored for the affective facilitation effect was that if affective valence influences consolidation of information transferred into WM it may facilitate the consolidation conditionally, instead of interfering with it. Given facilitating the consolidation, it should be reflected in WM performance. Regarding consolidation, it is assumed that transient perceptual representations are consolidated into a more durable form of WM by a process that takes time and has limited capacity (Jolicoeur and Dell'Acqua, 1998; Potter, 1976; Vogel, Luck, and Shapiro, 1998; Vogel, Woodman, and Luck, 2006).

To examine this, the research used an affective retro-cues memory task. In a retro-cue paradigm, when visual stimuli are presented, the perceptual representations should decay in the interval between the offset of the stimuli and the onset of the cue (that retrospectively selects the to-be-remembered [TBR] item). The selected representation should then be maintained in WM. We expected that when a negative word was used as a retro-cue, it would attract more attention to the TBR item, compared with a neutral or positive cue, resulting in a more available WM representation. Thus, performance on a speeded WM task (i.e., retro-cue memory task) should be better for items cued with negative words rather than with neutral or positive words.

However, negative valence makes it difficult to disengage attention from the words. This attention dwell effect for negative stimuli has been revealed even for non-anxious individuals (e.g., Gotoh, 2008; Gotoh et al., 2008). Accordingly, it is difficulty to disengage from the negative words as cues and this is expected to impair subsequent processing of the TBR information selected by the cue. Thus, we had to remove the difficulty of disengagement from the negative valence as a retro-cue. To remove it of negative valence, a color discrimination task that measured WM performance, using affective words as retro-cues for color discrimination, was combined with a target detection task that manipulated attention disengagement. By inserting the target detection task after the presented cue, attention was either kept on or away from the cued place. This manipulation allows us to examine the facilitation effect and attention dwell effect separately. We expected that the facilitation effect may emerge on the condition in which attention was away from the cued place, whereas the attention dwell effect may emerge on the condition in which attention was kept on the cued place.

Consequently, on each trial four colors were displayed, each at a designated position. The colors disappeared and an affective or neutral cue appeared in one of the positions, selecting that color as the TBR item. Next, the target detection task was inserted to either maintain spatial attention to the same location or shift it to a different location. Finally, participants were required to discriminate the TBR color from two distractors. Thus, this procedure makes it possible to separate the influences of negative affect in terms of attention dwell and the possible facilitation effect.

The results provided evidence that negative words influenced WM performance when used as attentional cues, indicating the affective interference effect as well as an affective facilitation effect. Regarding the interference effect, when attention was kept on the same location as the negative cue, a subsequent color discrimination task increased the RT. In contrast, regarding the facilitation effect, when attention was diverted away from the negative cue, the subsequent color discrimination task decreased (i.e., facilitated) the RT. The results suggest that the facilitation effect of negative valence may be due to more available representations in WM, capturing more attention by negative cue relative to neutral or positive cues.

The results suggest that the facilitation effect of negative cues can be explained in terms of an influence of short-term consolidation (Jolicoeur and Dell'Aqua, 1998; Potter, 1976; Vogel et al., 1998, 2006). Negative affective valence may contribute to the consolidation process by attracting attention to TBR items, resulting in more available WM representations, and improving WM performance. If a more available representation is not postulated, it would be difficult to explain why negative cues facilitated WM performance when attention was shifted away from the cued location. Since selective attention serves as a rehearsal mechanism for WM (Awh, Jonides, and Reuter-Lorenz, 1998), the attention shift usually could be thought to play into reduced performance on WM compared with attention dwell. Consistent with this, we observed shorter color discrimination RT when the target appeared in the same location as the cue and the cue was positive or neutral.

Conclusions for the Affective Facilitation Effect on Visual WM

The results in a retro-cue memory task successfully showed both interference and facilitation effects within the same task, manipulating the attention condition on the affective words. Firstly, thus, the result replicated the affective interference effect on WM with negative words, secondly, the results revealed the affective facilitation effect with negative words. The results suggest that the affective interference effect may be due to the difficulty of the attention shift (i.e., attention dwell) from the negative words, whereas the attention attraction for the negative words may cause the affective facilitation effect if the attention dwell is carried away from the words. We proposed that the facilitation effect may be due to the short-term consolidation of WM.

Regarding negative and positive affective words, at least in my series of experiments, only negative words caused the interference effect in an updating task and a retro-cue memory task, while both negative and positive words caused it in a counting task. Because arousal was controlled for between negative and positive valence, the same effect of negative and positive words would be expected due to the arousal or to the relationship between valence and arousal. In visual WM, negative words consistently influenced WM processing and caused the affective interference effect. In addition, only the negative words caused the affective facilitation effect in a retro-cue memory task. Taking these results, the influence of affective valence on WM may not be due to arousal, but due to affective valence.

INFLUENCE OF AFFECTIVE VALENCE ON AUDITORY WORKING MEMORY

In visual WM, both affective interference and facilitation effects were shown. Next, an attempt was made to reveal the impact of affective words on auditory WM. In everyday life, we are exposed to auditory affective words as well as to visual affective words. Do auditory affective words influence WM processing? To answer this question I explored the impact of affective valence on auditory WM using auditory affective words (Gotoh, 2012). Like in visual WM, the performance on an auditory WM task was compared between affective words (negative and positive words) and neutral words. Results showed both affective "pop up" or dominance effects and interference effects.

The Dominance and Interference Effects on Auditory WM: Evidence from an Auditory Recall Task (Gotoh, 2012)

According to Cowan (1984, 1999), a short auditory store corresponds to ongoing neural activity for 200-300 ms, and a long auditory store holds information for seconds involved in WM. Like in the visual WM study, I used auditory affective words in an auditory WM task. I had conducted 6 experiments on the influence of affective words on auditory WM (Gotoh, 2009). Among them, the present study focused on investigating the affective interference effect on auditory processing, including the dominance effect. That is, the present study investigated whether the affective valence of words interferes with ongoing auditory processing (after their recall) due to attention dwell on the affective words in auditory WM. Drawing a direct line to the results on affective interference in visual WM, if affective auditory words capture attention and influence the disengagement component of WM, the attention dwell would reduce subsequent recall on WM.

To investigate the influence on auditory WM processing, on each trial, negative, positive, and neutral words were simultaneously presented by three different speakers (left 45°, center, right 45°), using different spoken words with different women's voices. All words were composed of three mora (e.g., ko/ku/go) and they had the accent on the second mora. The words were digitally started simultaneously and the accent of the three words was at the same pitch. Speaker position and word valence were balanced. Participants were asked to recall the words right after they had been presented. Importantly, participants were asked to recall the words presented from the center speaker first and then to recall the words presented from both side speakers. This procedure allowed me to investigate the affective interference effect on auditory WM. Given attentional dwell on emotional auditory words, the recall of affective words (relative to neutral words) from the center speaker should reduce subsequent recall from the side speakers. In addition, the first recall from the center speaker could be expected to show how auditory WM processes auditory information when it arrives simultaneously. Participants were also asked to identify the speaker when they recalled a word to evaluate recognition of the sound orientation of recalled words. Moreover, as with visual WM, an anxiety inventory was administered to assess the influence of anxiety on the emotional interference effect (Shimizu and Imae, 1981; Spielberger, Gorsuch, and Lushene, 1970).

Regarding first recall, negative and positive words were more frequently recalled from the center speaker, and recall of negative and positive words from the center speaker reduced subsequent recall of other words from the side speaker. Accordingly, negative and positive words were dominantly recalled from the center speaker and the recall of negative and positive words interfered with ongoing auditory WM due to attention dwell, and consequently fewer words were subsequently recalled. Thus the results showed the dominance effect and interference effect of auditory affective words. Like for visual WM, the dominance and interference effects were not associated with the participants' level of anxiety. This suggests that both affective influences may be due to a central attention mechanism on audition rather than to an attention bias from anxiety. The dominance effect of affective words was supported by other experiments (Gotoh, 2009).

Reduced subsequent recall of other words emerged especially on the left speaker. On the right speaker, no affective interference emerged. Therefore, the results suggest that the right hemisphere is involved in the influence of affective valence on auditory WM and the right hemisphere may be involved in the attentional dwell effect of affective valence on auditory WM. This interpretation is consistent with earlier studies suggesting that there is hemispheric lateralization for emotional content and that the right hemisphere is dominant for the emotional effect (e.g., Morris, Scott, and Dolan, 1999; Pinel, 2003).

Conclusions for the Dominance and Interference Effects on Auditory WM

The human auditory information processing system has a remarkable attention mechanism for processing linguistic information. For instance, considerable evidence exists that people can easily filter one speaker's message from several other speakers' messages competing for our attention (Cherry, 1953; Conway, Cowan, and Bunting, 2001; Giard, Fort, Mouchetant-Rostaing, and Pernier, 2000). This well-known "cocktail party effect" indicates that the auditory system has a remarkable attention mechanism to process linguistic information.

In addition to auditory ability, the present study (Gotoh, 2012) revealed that auditory simultaneous message processing is strongly influenced by affective valence. Apparently, in auditory information processing all messages are not processed equally regardless of affective valence of the words: the observed affective influence appeared in the form of dominance and interference effects. Negative and positive words were dominantly recalled relative to neutral words at first and then, the recall of affective words reduced the subsequent recall of other words, interfering with auditory WM processing.

Conclusion

My series of experiments attempted to reveal whether affective words influence visual and auditory WM processing. The impact of affective words was shown on both visual and auditory WM processing. In visual WM, negative, or negative and positive words induced an affective interference effect, in which affective words increased RT or error rates, reducing speed or accuracy of WM processing (Gotoh, 2008; Gotoh, et al., 2008; Gotoh et al., 2010). Moreover, negative words induced a facilitation effect, in which negative words decreased RT, facilitating speed of WM processing (Gotoh et al., 2010). In auditory WM, negative and positive words induced a dominance effect, in which affective words "popped up" from the center speaker. Then, after the dominant recall, negative and positive words induced an interference effect, in which the affective words reduced subsequent other recall (Gotoh, 2012). These affective influences were not due to an attention bias from anxiety, but due to the central attention mechanism of WM, regardless of the visual or auditory domain. From these results, I conclude that the WM processing system is modulated for affective valence of perceived information, regardless of the domain. In other words, human working spaces of vision and audition do not process incoming information equally and constantly, but are influenced by the affective valence of that information, resulting in changes of the sequence, speed, and accuracy of information processing. These affective influences suggest that since our central information system is of limited capacity (e.g., Cowan, 1999, 2005), important information should be selected from other information. Affective modulation may include neuromodulation (e.g., Anderson and Phelps, 2001; Martino, Kalisch, Rees, and Dolan, 2009).

Drawing a direct line to the visual WM results, the possible factor that distinguishes between affective interference and facilitation effects may be attention dwell on affective words. As shown by Gotoh et al., (2010), negative words induced an interference effect when attention stayed on negative words. However, in eliminating that attention dwell on negative words, the latter induced affective facilitation, to which more available WM representation may have contributed. Thus, attention dwell could induce affective interference by hampering on-going WM processing. However, there is also the possibility that eliminating the dwell may reveal affective facilitation. Obviously, if the procedure did not eliminate attention dwell from the affective words, facilitation would not emerge; it may have been covered by an interference effect.

In order to explain the facilitation effect, it may be possible that when affective words come in, they may capture central WM attention first, while the attention capture may enhance, for instance, the short-term consolidation in WM. Then, if attention stays on the affective words continuously, it may contribute to an interference effect, interfering with ongoing processing. In contrast, if attention is diverted away from the affective words after the attention capture, the enhanced WM representation may contribute to the facilitation effect. Thus, I propose that affective words may capture attention of WM first when affective information comes in, and then attention may generally be difficult to disengage from the words and dwell on the words continuously if attention is not diverted. In line with this, findings on the affective interference effect (Gotoh, 2008; Gotoh et al., 2008) indicate that mental manipulation with affective words may generally cause attention interference. I will explain more about this flow in the next paragraph, in terms of attention capture and dwell for the corresponding affective dominance, and interference and facilitation effects, respectively.

The attention capture for incoming affective information may cause "pop up" or dominance effects in vision and audition. In the aforementioned series of experiments, affective dominance and interference effects were shown only in auditory WM whereas interference and facilitation effects were shown in visual WM. Therefore, affective dominance was only shown in auditory WM but not in visual WM in the experiments. However, previous visual studies have found that negative words enter consciousness faster than neutral words (Anderson and Phelps, 2001; Kihara and Osaka, 2008; Martino et al., 2009). These studies used a rapid serial visual presentation (RSVP) task that is thought to be a WM task (Martino et al., 2009; Smith and Jonides, 1999; Todd and Marois, 2004). For instance, Kihara and Osaka (2008) showed that negative words reduced attentional blink in RSVP tasks, suggesting that negative words come into consciousness faster than neutral words. Obviously, these results seem to indicate "pop up" or dominance effects of affective words in visual WM. Thus, the result on the RSVP task and affective "pop up" recall in my auditory experiment could indicate dominance effects of affective words on visual and auditory WM, respectively. Accordingly, affective words (negative and positive words), at least negative words, may be encoded faster and enter consciousness faster than neutral words, resulting in "pop up" or dominance effects, regardless of the domain.

After the dominance of information processing of negative, or negative and positive words, the attention condition may generate interference or facilitation effects. If attention dwells on the affective words, since attention is grabbed by affective words it interferes with ongoing central processing, resulting in interference. In contrast, if attention is diverted away from the affective words, it may reveal, for instance, enhanced WM representations, resulting in a facilitation effect. In this way, when affective words come in, attention capture may invoke affective "pop up" or dominance first, and then attention dwell may invoke affective interference, regardless of visual or auditory WM. In both domains since affective words are deemed important, they may reach consciousness faster and cause interference with the central executive.

Regarding affective valence, negative and positive valence did not have the same impact on WM processing. In visual WM, negative words always caused interference and facilitation in the three tasks, whereas positive words did not always cause interference and never caused facilitation. In auditory WM, negative and positive words caused dominance effects and they caused interference effects. As mentioned above, this negative-positive difference could be explained by a negativity bias assuming that negative, but not positive stimuli interrupt processing due to their relevance for survival (e.g., Pratto and John, 1991). Positive valence might impinge more slowly than negative valence. For instance, Maljkovic and Martini (2005) found that positive and negative scenes differently influenced short-term memory. In their study, positive valence scenes accumulated memory at a constant rate whereas negative scenes were encoded slowly at first and then increasingly faster.

An important topic that I should mention is the relationship between valence and arousal. Other studies also have found the influence of valence and arousal in cognition. Some studies found an influence of valence but not of arousal (Kihara and Osaka, 2008), whereas other studies found the influence of arousal but not of valence (Anderson, 2005; Mather, Mitchell, Raye, Novak, Greene and Johnson, 2006). In addition, some studies found the effects caused by the relationship between valence and arousal (Verbruggen and de Houwer, 2007). As mentioned above, our results revealed that the negative and positive words had different impacts on WM. In my experiments, since arousal was matched between negative and positive words, if arousal is crucial for that influence, negative and positive words should show the same impact, compared to neutral words. However, the two valences showed different impacts on processing. Thus, the affective dominance, interference, and facilitation effects could be due to affective valence. Although the results suggest this conclusion, further research is needed to reveal the relationship between valence and arousal.

ACKNOWLEDGMENTS

I wish to thank Dr. Christian Robnagel for checking the English.

REFERENCES

Anderson, A. K. (2005). Affective influences on the attentional dynamics supporting awareness. *Journal of Experimental Psychology: General, 134*, 258-281.

Anderson, A. K., and Phelps, E. A. (2001). Lesions of the human amygdala impair enhanced perception of emotionally salient events. *Nature, 411*, 305-309.

Awh, E., Jonides, J., and Reuter-Lorenz, P. A. (1998). Rehearsal in spatial working memory. *Journal of Experimental Psychology: Human Perception and Performance, 24*, 780-790.

Baddeley, A. and Logie, R. (1999). Working memory: the Multiple-component model. In A. Miyake and P. Shah (Eds.), *Models of working memory: Mechanisms of active maintenance and executive control* (pp.28-61). Cambridge, UK: Cambridge University Press.

Berthoz, S., Blair, R. J. R., Le Clec'h, G., and Martinot, J. L. (2002). Emotions: From neuropsychology to functional imaging. *International Journal of Psychology, 37*, 193–203.

Braver, T. S., Cohen, J. D., Nystrom, L.E., Jonides, J., Smith, E. E., and Noll, D. C. (1997). A parametric study of prefrontal cortex involvement in human working memory. *Neuroimage, 5*, 49-62.

Cherry, E. C. (1953). Some experiments on the recognition of speech, with one and with two ears. *Journal of the Acoustical Society of America, 25*, 975-979.

Christianson, S. A. (Ed.) (1992). *The handbook of emotion and memory: Research and theory*. Hillsdale, NJ: Lawrence Erlbaum Associates Inc.

Cohen, J. D., Perlstein, W. M., Braver, T. S., Nystrom, L. E., Noll, D. C., Jonides, J., and Smith, E.E.(1997). Temporal dynamics of brain activation during a working memory task. *Nature, 386*, 604-608.

Colombel, F. (2000). The processing of emotionally positive words according to the amount of accessible retrieval cues. *International Journal of Psychology, 35*, 279–286.

Conway, A. R. A., Cowan, N., and Bunting, M. F. (2001). The cocktail party phenomenon revisited: The importance of working memory capacity. *Psychonomic Bulletin and Review, 8*, 331-335.

Conway, A. R. A., Cowan, N., Bunting, M. F., Therriault, D. J., and Minkoff, S. R. B. (2002). A latent variable analysis of working memory capacity, short-term memory capacity, processing speed, and general fluid intelligence. *Intelligence, 30*, 163-184.

Cowan, N. (1984). On short and long auditory stores. *Psychological Bulletin, 96*, 341-370.

Cowan, N. (1999). An embedded-processes model of working memory. In A. Miyake and P. Shah (Eds.), *Models of working memory: Mechanisms of active maintenance and executive control* (pp.62-101). Cambridge, UK: Cambridge University Press.

Cowan, N. (2001). The magical number 4 in short-term memory: A reconsideration of mental storage capacity. *Behavioral and Brain Sciences, 24*, 87-185.

Cowan, N. (2005). Working memory capacity. New York: Psychology Press.

Daneman, M., and Carpenter, P. A. (1980). Individual differences in working memory and reading. *Journal of Verbal Learning and Verbal Behavior, 19*, 450-466.

Daneman, M., and Merikle, P. M. (1996). Working memory and language comprehension: A meta-analysis. *Psychonomic Bulletin and Review, 3*, 422-433.

Doerksen, S., and Shimamura, A. P. (2001). Source memory enhancement for emotional words. *Emotion, 1*, 5–11.

Engle, R. W., Tuholski, S. W., Laughlin, J. E., and Conway, A. R. A. (1999). Working memory, short-term memory, and general fluid intelligence: A latent-variable approach. *Journal of Experimental Psychology: General, 128*, 309-331.

Fox, E., Russo, R., Bowles, R., and Dutton, K. (2001). Do threatening stimuli draw or hold visual attention in subclinical anxiety? *Journal of Experimental Psychology: General, 130*, 681–700.

Giard, M. H., Fort, A., Mouchetant-Rostaing, Y., and Pernier, J. (2000). Neurophysiological mechanisms of auditory selective attention in humans. *Frontiers in Bioscience, 5*, d84-d94.

Gotoh, F. (2008). Influence of affective valence on working memory processes. *International Journal of Psychology, 43*, 59-71.

Gotoh, F. (2009). *Influence of the affective valence of words on the processing in working memory.* Unpublished doctoral dissertation, University of Tsukuba, Japan.

Gotoh, F. (2012). Affective valence of words impacts recall from auditory working memory. *Journal of Cognitive Psychology, 24*, 117-124.

Gotoh, F. and Ohta, N. (2001). Affective valence of two-compound Kanji words. *Tsukuba Psychology Research, 23*, 45–52. (in Japanese)

Gotoh, F., Kikuchi, T., and Roßnagel, C. S. (2008). Emotional interference in enumeration: A working memory perspective. *Psychology Science Quarterly, 50*, 526-537.

Gotoh, F., Kikuchi, T., and Olofsson, U. (2010). A facilitative effect of negative affective valence on working memory. *Scandinavian Journal of Psychology, 51*, 185-191.

Hamann, S. (2001). Cognitive and neural mechanisms of emotional memory. *Trends in Cognitive Sciences, 5*, 394–400.

Jolicoeur, P., and Dell'Acqua, R. (1998). The demonstration of short-term consolidation. *Cognitive Psychology, 36*, 138-202.

Kihara, K., and Osaka, N. (2008). Early mechanism of negativity bias: An attentional blink study. *Japanese Psychological Research, 50*, 1-11.

King, J., and Just, M. A. (1991). Individual differences in syntactic processing: The role of working memory. *Journal of Memory and Language, 30*, 580-602.

Kyllonen, P. C., and Christal, R. E. (1990). Reasoning ability is (little more than) working memory capacity?! *Intelligence, 14*, 389-433.

MacDonald, M. C., Just, M. A., and Carpenter, P. A. (1992). Working memory constraints on the processing of syntactic ambiguity. *Cognitive Psychology, 24*, 56-98.

Mandler, G., and Shebo, B. J. (1982). Subitizing: An analysis of its component processes. *Journal of Experimental Psychology: General, 111*, 1-22.

De Martino, B., Kalisch, R., Rees, G., and Dolan, R. J. (2009). Enhanced processing of threat stimuli under limited attentional resources. *Cerebral Cortex, 19*, 127-133.

Maljkovic, V., and Martini, P. (2005). Short-term memory for scenes with affective content. *Journal of Vision, 5*, 215–229.

Mather, M., Mitchell, K. J., Raye, C. L., Novak, D. L., Greene, E. J., and Johnson, M. K. (2006). Emotional arousal can impair feature binding in working memory. *Journal of Cognitive Neuroscience, 18*, 614-625.

McNally, R. J., Riemann, B. C., and Kim, E. (1990). Selective processing of threat cues in panic disorder. *Behavior Research and Therapy, 28*, 407–412.

Miller, G. A. (1956). The magical number seven, plus or minus two: Some limits on our capacity for processing information. *Psychological Review, 63*, 81-97.

Miyake. A and Shah. P (Eds.) (1999). *Models of working memory: Mechanisms of active maintenance and executive control*. Cambridge, UK: Cambridge University Press.

Morris, J. S., Scott, S. K., and Dolan, R. J. (1999). Saying it with feeling: Neural responses to emotional vocalizations. *Neuropsychologia, 37*, 1155-1163.

Oberauer, K. (2002). Access to information in working memory: Exploring the focus of attention. *Journal of Experimental Psychology: Learning, Memory, and Cognition, 28*, 411-421.

Pinel, J. P. J. (2003). Lateralization, language, and the split brain. In P. John (Ed.), *Biopsychology* (pp. 406-436). Cambridge, UK: Cambridge University Press.

Potter, M. C. (1976). Short-term conceptual memory for pictures. *Journal of Experimental Psychology: Human Learning and Memory, 2*, 509-522.

Pratto, F., and John, O. P. (1991). Automatic vigilance: The attention-grabbing power of negative social information. *Journal of Personality and Social Psychology, 61*, 380–391.

Reisberg, D., and Hertel, P. (Eds.) (2004). *Memory and emotion*. Oxford, UK: Oxford University Press.

Salthouse, T. A., Babcock, R. L., and Shaw, R. J. (1991). Effects of adult age on structural and operational capacities in working memory. *Psychology and Aging, 6*, 118–127.

Segal, Z. V., Gemar, M., Truchon, C., Guirguis, M., and Horowitz, L. M. (1995). A priming methodology for studying self-representation in major depressive disorder. *Journal of Abnormal Psychology, 104*, 205–213.

Shimizu, H., and Imae, K. (1981). Development of the Japanese edition of the Spielberger State Trait Anxiety Inventory (STAI) for student use. *Japanese Journal of Educational Psychology, 29*, 348-353.

Smith, E. E., and Jonides, J. (1999). Storage and executive processes in the frontal lobes. *Science, 283*, 1657-1661.

Smith, E. E., Jonides, J., and Koeppe, R. A. (1996). Dissociating verbal and spatial working memory using PET. *Cerebral Cortex, 6*, 11-20.

Spielberger, C. D., Gorsuch, R. L., and Lushene, R. E. (1970). *Manual for the State–Trait Anxiety Inventory (Self-Evaluation Questionnaire)*. Palo Alto, CA: Consulting Psychologists Press.

Todd, J. J., and Marois, R. (2004). Capacity limit of visual short-term memory in human posterior parietal cortex. *Nature, 428*, 751-754.

Verbruggen, F., and De Houwer, J. (2007). Do emotional stimuli interfere with response inhibition? Evidence from the stop signal paradigm. *Cognition and Emotion, 21*, 391-403.

Vogel, E. K., Luck, S. J., and Shapiro, K. L. (1998). Electrophysiological evidence for a postperceptual locus of suppression during the attentional blink. *Journal of Experimental Psychology: Human Perception and Performance, 24*, 1656-1674.

Vogel, E. K., Woodman, G. F., and Luck, S. J. (2006). The time course of consolidation in visual working memory. *Journal of Experimental Psychology: Human Perception and Performance, 32*, 1436-1451.

Williams, J. M. G., Mathews, A., and MacLeod, C. (1996). The emotional Stroop task and psychopathology. *Psychological Bulletin, 120*, 3–24.

Yiend, J., and Mathews, A. (2001). Anxiety and attention to threatening pictures. *The Quarterly Journal of Experimental Psychology, 54A*, 665–681.

In: Working Memory
Editor: Helen St. Clair-Thompson

ISBN: 978-1-62618-927-0
© 2013 Nova Science Publishers, Inc.

Chapter 9

VERBAL AND VISUO-SPATIAL PROCESSES IN SPATIAL ORIENTATION AND NAVIGATION

Andre Garcia and Carryl L. Baldwin
George Mason University, Fairfax, VA, US

ABSTRACT

Processing spatial information is a common and important task of everyday life. Auditory information serves to help us quickly orient to important objects and events in the world and verbal directions or visuospatial maps help us learn to navigate through and familiarize ourselves with new areas. Recent evidence from our labs indicates that these two types of spatial tasks – basic orienting and navigation share important mechanisms. Further, these mechanisms are strongly influenced by individual differences. In a series of studies ranging from learning and remembering novel environments to handling conflict in basic spatial orientation tasks we show that some individuals inherently process verbal information faster and more efficiently while others process visuospatial information better. In this chapter we elaborate on these new discoveries and discuss their implications for working memory, developmental differences and spatial cognition in everyday life.

INTRODUCTION

Processing spatial information and learning to navigate through the environment are critical tasks in the everyday life of most humans. These tasks require attention and there are large differences across individuals in how well these tasks can be performed alone and in combination with other tasks (Baldwin and Reagan, 2009). In this chapter we discuss some primary aspects of verbal and visuospatial processing as they relate to spatial orientation and navigation. Throughout our discussion emphasis will be placed on the working memory mechanisms at play and individual and developmental differences in these abilities. We begin by defining what we mean by spatial orientation and navigation and then proceed to briefly discuss some major paradigms that we have used to investigate these issues followed by a review of relevant literature both from our lab and from other published work.

DEFINING SPATIAL ORIENTATION AND NAVIGATION

Spatial and navigational abilities are multidimensional and have been assessed with a wide variety of tasks (Coluccia and Louse, 2004). The wide variability in tasks utilized by various researchers is no doubt at least partially responsible for the disparity among results. Therefore, a few definitions are warranted at the outset. Loomis, Klatzky, Golledge, and Philbeck (1999) define navigation as the planning of travel through the environment, updating position and orientation during travel, and in the event of becoming lost, reorienting and reestablishing travel toward the destination. More simply put, it is the aggregate task of wayfinding and motion. Wayfinding primarily refers to making ones way from one point to another. It does not necessarily involve any attempt to remember or learn either the route taken or the relationship of objects and landmarks in the environment. Spatial orientation is the process of calculating your heading and position in relation to landmarks from an egocentric perspective and/or in relation to a larger geocentric spatial framework (Klatzky, Loomis, Beall, Chance, and Golledge, 1998). Learning the relationship between major landmarks and geographical features can be defined as developing a cognitive map and is referred to as the development of survey knowledge (Thorndyke and Hayes-Roth, 1982). Individuals differ in each of these aspects of navigation and will tend to rely on different amounts and types of working memory resources depending on their inherent abilities and the strategies they use.

INDIVIDUAL DIFFERENCES IN WAYFINDING

Considerable evidence now indicates that people develop different strategies for wayfinding and navigation that are accompanied by a tendency to orient themselves in reference to either egocentric (a first-person, field forward perspective) or geocentric perspective, which is also called allocentric and is based on cardinal directions irrespective of an individual's current orientation (Baldwin, 2009; Baldwin and Reagan, 2009; Hegarty, Montello, Richardson, Ishikawa, and Lovelace, 2006; Kato and Takeuchi, 2003; Lawton, 2001; Saucier, Bowman, and Elias, 2003). People who wayfind from an egocentric perspective will tend to need directions to go left or right at particular landmarks and will follow a verbal, sequential set of directions. They will typically tend to memorize routes by these sequential verbal lists rather than forming a cognitive map or picture of the environment in their head and they are typically not good at using cardinal directions (Baldwin and Reagan, 2009; MacFadden, Elias, and Saucier, 2003; Pazzaglia and De Beni, 2001).

People navigating from a geocentric perspective will be able to keep track of the relationships between major landmarks and maintain awareness of their cardinal heading. They will be able to use directions such as go north of the city center five blocks and then turn east and continue another six blocks and the destination will be on the south side of the library. People who are able to navigate from a geocentric perspective tend to form a cognitive map of the area they are traversing and use visuospatial working memory resources in this process (Baldwin and Reagan, 2009; Cornoldi and Vecchi, 2003; Garden, Cornoldi, and Logie, 2002). Another important distinction is that people who navigate from a geocentric perspective tend to be able to use an egocentric perspective if the situation calls for it. But the reverse is not true. That is, people who tend to rely on an egocentric, verbal-sequential strategy tend to have difficulty if not find it impossible to use a geocentric perspective (Gugerty and Brooks, 2004; Saucier et al., 2003). There has been a slight but consistent trend for males to prefer a geocentric, visuospatial perspective while females have a tendency to prefer the egocentric, verbal-sequential strategy (Lawton, 1994; Saucier et al., 2003). However, as will be discussed, these tendencies are far from clear cut and the sex-based distinction appears to be lessening over time. These individual differences and their associated working memory mechanisms will be covered to a greater extent later in this chapter.

WORKING MEMORY IN SPATIAL NAVIGATION

Navigational Strategies and Spatial Representations

How humans navigate and process spatial information is of critical importance. As previously discussed, there are two primary ways of representing spatial information about the environment. These are egocentric and allocentric (or geocentric) representations (Blajenkova, Motes, and Kozhevnikov, 2005; Lawton, 1994). Egocentric, literally meaning in relation to the person, is a first-person forward-up view, indicating what is in front of the user. Allocentric or geocentric, meaning in relation to the earth but can also be in relation to something else like a building, is generally a north-up point of view. These two distinct perspectives can be linked to a hierarchy of types of spatial information. Humans can encode spatial information as either landmark, route, or survey knowledge (Thorndyke and Hayes-Roth, 1982).

Previously it was thought that spatial information was acquired in a linear progression from landmark to route to survey knowledge. Thorndyke and Hayes-Roth (1982) thought that a person first gathers spatial information about the environment from orienting him or herself to landmarks. This provides predominantly egocentric reference information. As greater familiarity with the environment is acquired, route knowledge is developed in which more information is learned about the environment by learning a sequence of turns from taking an egocentric perspective at key landmarks. Finally, with even more experience with the environment, people are able to create a cognitive map of the area by connecting the relative location of landmarks to each, acquiring what is referred to as survey knowledge. Survey knowledge results in a cognitive map and understanding the spatial location of landmarks from a geocentric perspective. It is now understood that the fixed progression from landmark to route to survey knowledge is largely not necessary. It is possible to learn survey-like geocentric information from a single exposure and it is also possible that certain individuals are unable to learn or use survey knowledge and can only use egocentric or verbal information (Blajenkova et al., 2005). Further, even with considerable experience with a given environment, some people continue to have great difficulty acquiring survey knowledge.

From maps, some people may acquire survey knowledge about the area, which is generally allocentric in nature (Thorndyke and Hayes-Roth, 1982). Further, it is possible for some individuals with high spatial abilities to gain survey knowledge after just one in-person exposure to a novel environment even without the ability to study a map (Blajenkova et al., 2005). But those who tend to rely on an egocentric perspective will have great difficulty forming survey knowledge and may be unable to benefit from map usage.

A geocentric north-up map is generally considered a more robust form of representing spatial information, relative to an egocentric. However, there are some limitations to geocentric maps. The primary one is that not all individuals will be able to effectively use them and thus they may increase the probability of people getting disoriented or lost. Even if they can be used, they generally require greater mental effort since they generally require mental rotation due to reference frame misalignment (Gugerty and Brooks, 2001), having to transform the geocentric information from the map to usable egocentric information for navigation purposes (Gugerty and Brooks, 2004). The reference frame misalignment may be the most critical flaw of survey knowledge gained from a north-up map. If an environment is learned from a north-up perspective, but navigation begins from a different entry point, the spatial encoding of landmarks and relationships between landmarks and the environment may seem off because the geocentric orientation-specific representation of the environment (Richardson, Montello, Hegarty, and 1999) It is also difficult to represent heading information in an intuitive manner in geocentric survey style maps (Gugerty and Brooks, 2001).

Conversely, an egocentric perspective gives a forward-up representation of the environment, showing the user what is directly in front of him or her and when salient landmarks are pointed out they facilitate the development of route knowledge (Reagan and Baldwin, 2006).

Egocentric maps are generally easy for navigators of both styles to use for basic wayfinding, but may fall short of assisting the user in forming survey knowledge. Egocentric based navigation aids are quick to learn how to use (Van Erp, Van Veen, and Jansen, 2006) generally considered to require less workload to use, and elicit faster responses (Streeter, Vitello, and Wonsiewicz, 1985; Van Erp and Van Veen, 2004), and are an effective method of route guidance, even when they are not the preferred format of the user (Furukawa, Baldwin, and Carpenter, 2004).

One particular advantage of egocentric navigation aids is that other non-visual modalities lend themselves well to egocentric directions. Whether it is an auditory route guidance system (Baldwin, 2009), or tactile navigation aid (Van Erp et al., 2006), presentation of route information in non-visual modalities is a lot more conducive to egocentric formats than geocentric displays.

However, Baldwin and Reagan (2006) demonstrated that both formats could be used successfully to facilitate route learning through auditory guidance. Next we turn attention to a more focused account of sources of individual differences in spatial and navigation ability pointing out the role of working memory as applicable. But first, a brief review of the dominant model of working memory is warranted.

Verbal and Visuospatial Working Memory

Baddeley and Hitch (1974) described a model of working memory that has dominated cognitive psychology for many years. In their early tri-part model, a central executive component controlled and allocated available resources to two slave systems, the phonological loop and the visuospatial sketchpad. As the names suggest, the phonological loop was thought to be used to process auditory-verbal information while the visuospatial sketchpad was specialized in processing visual and spatial information (Baddeley, 1992; Baddeley and Hitch, 1974; Baddeley and Hitch, 1994). In later years, Baddeley (see Baddeley, 2002; Baddeley, 2003; Baddeley, 2000) added a fourth component, which he called the episodic buffer. The episodic buffer is thought to aid the limited capacity central executive, particularly in coordinating the integration of information between working and long term memory. Baddeley's model of working memory has been extremely influential over the ensuing years and continues to play an important role in explaining differences in working memory processes used by individuals with different navigational strategies.

Working Memory during Navigation

Numerous factors influence spatial abilities and the working memory required during navigational tasks. Sex, experience, age, navigational strategy, and working memory capacity have all been found to account for significant amounts of variance in performance on navigational tasks (Baldwin and Reagan, 2009; Choi and Silverman, 2003; Cooper and Brooks, 2004; Gugerty and Brooks, 2004; Lawton, 1994). Further, even in the absence of observable performance differences individuals may exhibit different patterns of brain activation (Hugdahl, Thomsen, and Ersland, 2006) and different patterns of working memory interference (Baldwin and Reagan, 2009) that are indicative of the use of different processing strategies. Differences in navigation abilities based on sex are one of the most cited forms individual differences.

Sex Differences

Sex differences in spatial task performance, with males outperforming females are a relatively consistent finding (Coluccia, Bosco, and Brandimonte, 2007; Coluccia and Louse, 2004). In addition to different patterns of performance, males and females tend to use different strategies. Lawton (1994) found that females tended to use route-based methods whereas males tended to use orientation methods where they maintain their frame of reference with the environment. Women also reported higher subjective anxiety due to navigation, and that spatial perception abilities had a positive correlation with the orientation strategy, suggesting that those who tend to maintain their geocentric orientation tend to have higher spatial perception ability. Saucier et al. (2003) found that women tended to use landmark based strategies more often than men, and that they were better at remembering landmarks. Regardless of sex, differences in sense of direction also play a large role in navigational ability.

Sense of Direction

Furthermore, an individual's self-reported sense of direction (SOD) ability can generally predict navigation performance as well as strategy. Kato and Takeuchi (2003) found that individuals with a self-reported good sense of direction (GSD) tend to maintain their heading in relation to cardinal positions. Conversely, individuals reporting a poor sense of direction (PSD) experience great difficulty maintaining their cardinal heading. PSD individuals are more likely to use ineffective strategies and have a more difficult time navigating and maintaining awareness of their orientation. Thus, individuals with PSD tend to use more working memory resources but do so ineffectively. Baldwin (2009) reports additional differences between GSD and PSD individuals. Specifically, PSD individuals tend to prefer egocentric auditory based route guidance systems that provide point by point navigational guidance. Conversely, those with a GSD tend to prefer visual based geocentric route guidance systems in the form of a map showing not only point by point guidance but a general overview of the area to be traversed.

Furukawa et al. (2004) elaborated on the relationship between individual differences in SOD and preference for route guidance format and modality and also perceived workload. Individuals with a PSD preferred auditory route guidance presented from an egocentric perspective. For example, egocentric directions consisted of statements such as, "Turn right in two blocks." People with a poor sense of direction preferred to be directed from place to place and were much less capable relative to their good sense of direction counterparts at answering cardinal heading information about routes they had traversed and at remembering the relationship between key landmarks encountered along the route. This was taken as evidence that an individual with a poor sense of direction has difficulty forming a cognitive map of areas he or she has traversed in a driving simulation.

Conversely, individuals with a GSD prefer a geocentric perspective, and tend to be able to use cardinal heading information and form a cognitive map (Furukawa et al., 2004; Kato and Takeuchi, 2003). Furukawa, et al., found that individuals with a GSD preferred a geocentric map over egocentric directions, although they were capable of using the egocentric directions if need be. Interestingly, individuals with a GSD had difficulty if both types of route guidance information were presented concurrently. Furukawa et al. interpreted this from the perspective that the auditory information would have been difficult to ignore and thus likely interfered with their use of their preferred strategy. Conversely, the PSD individuals would likely have found it easier to ignore the visual information during concurrent presentation. Therefore, the PSD performance was not disrupted when both types of information were presented. We examine more influences of these individual differences in navigational strategy next.

Working Memory Strategies

Fundamentally, individuals appear to differ in how they approach navigating through novel environments or performing spatial tasks and these differences rely on different working memory strategies. Baldwin and Reagan (2009) suggest that individuals with a poor sense of direction rely on verbal working memory when learning a route while navigating whereas those with a good sense of direction utilized visuospatial working memory. They assessed this by having participants learn a route while performing two different working memory interference tasks. One was an articulatory rehearsal task designed to interfere with verbal working memory (causing articulatory suppression). The other was a visuospatial tapping task designed to interfere with visuospatial working memory. Baldwin and Reagan (2009) found that those with a poor sense of direction had more difficulty learning routes while having to perform the concurrent verbal task, relative to when they had to learn routes while engaging in the visuospatial tapping task.

Conversely, those with a good sense of direction experienced more interference while attempting to learn a route while performing a concurrent visuospatial tapping task, relative to the verbal task, suggesting that they were relying on visuospatial working memory to perform the route learning task. Their strategy for navigation was competing with the resources concurrently required by the tapping task.

Thus, both tasks were fighting for visuospatial working memory resources at the same time. Similar results have been found previously in other labs (Saucier et al., 2003) and when attempting to learn novel routes in unfamiliar cities (Garden et al., 2002). In addition to these differences in type of working memory used as a function of the navigational strategy used, research suggests that those with a poor sense of direction work harder – use more working memory resources – to do the same task as their good sense of direction counterparts, even though they may not achieve the same performance levels.

Multimodal Way Point Navigation

Research from our lab further suggests that individual differences in multimodal waypoint navigation may also depend on display format (egocentric vs. geocentric) and modality. Garcia, Finomore, Burnett, Baldwin and Brill (2012) conducted a study to investigate individual differences in waypoint navigation via a visual, auditory, tactile, or multimodal route guidance system. Participants were lead via the various uni- or multi-modal route guidance system from waypoint to waypoint and were instructed to look for certain landmarks throughout the environment. Participants with a poor sense of direction revealed that those with a poor sense of direction were significantly slower than those with a good sense of direction in traversing the routes, particularly when using a traditional north-up egocentric map. Results of this experiment indicate that individual differences in waypoint navigation between those with a good sense of direction and those with a poor sense of direction are possible and display modality should be taken into consideration when designing navigation systems. Overall, the unimodal geocentric visual condition was the slowest and least accurate.

Those with a poor sense of direction had most difficulty with this condition and we believe this is partially due to the added workload required to perform mental rotations to align reference frame to the current view which is a task that those with a poor sense of direction may have difficulty with or are incapable of. These individual differences appear to manifest quite early in the developmental process.

DEVELOPMENTAL PATTERNS OF SPATIAL AND NAVIGATION ABILITIES

The development of different forms of spatial and navigational abilities has been of interest for some time (O'Keefe and Nadel, 1978; Thorndyke and Hayes-Roth, 1982; Tolman, 1924; Tolman, 1948). Here, we briefly review key developmental milestones for two major forms of spatial abilities - mental rotation and the allocentric representations.

Mental Rotation

The ability to mentally rotate objects is one form of spatial ability. Though its role in navigational abilities remains controversial, mental rotation is a frequently assessed spatial ability in both children and adults (Aretz and Wickens, 1992; Frick, Ferrara, and Newcombe, 2013; Hegarty et al., 2006). Recently, Frick et al. observed that children younger than age five were basically unable to perform a mental rotation task even when the figure or its mirror image were rotated by only slight angular disparities. Children at least five years of age were able to perform the mental rotation task, showing larger errors and slower response times as the angular disparity increased (Frick et al., 2013). Advanced age is associated with a decline in mental rotation abilities (Moffat, Zonderman, and Resnick, 2001). For example, Moffat et al. found that both mental rotation and the ability to locate a goal location in a virtual environment declined from age 22 to 91 years. The ability to form allocentric representations of an environment can be thought of as a higher order mental rotation ability.

Allocentric Representations

Developmental progress in the ability to form allocentric representations has been of interest for some time (for a review see Allen, 2004). Allocentric representations are a key component in the capability of map usage and thus it is an important spatial ability in navigation.

In young children it has often been examined by providing a small scale replica of a larger environment. A child is asked to watch as a small replica of a toy or other object is then placed in the small environment and then the child is asked to find the real object in the larger environment. If the child can understand the relationships between the two environments, he or she is thought to have mastered some level of allocentric representation (Allen, 2004; Ribordy, Jabès, Banta Lavenex, and Lavenex, 2013).

Using a variation of this technique, Ribordy et al. (2013) found that children between the age of 2 and 3 ½ years were unable to perform a task requiring them to locate three objects in a real environment after seeing their representations in the replicated environment. However, when the task was simplified so that only one object was to be found and it could only be located in one of four locations, children in this age group were generally able to successfully find the object. Children younger than age two were not able to perform even this simplified version of the task. Children over the age of 3 ½ years were generally able to successfully perform the more difficult three object task. These results indicate that the ability to form allocentric representations develops gradually with only rudimentary capabilities being present before 3 ½ years of age. We now turn attention to one final spatial ability, route learning.

Route Learning

Differences between males and females in spatial perception are present at least as early as age 9 years (Choi and Silverman, 2003) but differences in route learning appear later at around age 12 years (Choi and Silverman, 2003). Interestingly, differences between the route learning strategies used by males and females that emerge during adolescence seem to coincide with the working memory strengths of these individuals. Males tend to outperform females from a young age on tasks such as mental rotation but the results of a large meta-analysis indicate that these sex differences appear to be weakening over time (Voyer, Voyer, and Bryden, 1995). That is, the sex differences in spatial abilities found in many previous studies may be due in large part to differences in environmental exposure, with males generally having more opportunity and freedom to explore their environments than females. As parental and cultural practices change, giving females greater opportunities to explore their environments independently, the gap between male and female performance on spatial tasks appears to be decreasing.

As the gender gap in spatial abilities decreases greater focus can be placed on other sources of individual differences in navigational ability. The previously discussed strategy differences between use of egocentric verbal sequential information or geocentric visuospatial information is one such sex-independent source of difference (Baldwin and Reagan, 2009). Recent work from our lab suggests that these differences manifest themselves early in the processing stream and can be evidenced in basic laboratory tasks relying on semantic versus location spatial information.

Basic Process – Auditory Spatial Stroop

In order to examine basic spatial orientation behavior we developed a unique version of the Stroop task which we refer to as an Auditory Spatial Stroop paradigm (Barrow and Baldwin, 2010a). Recall that in the original Stroop paradigm (Stroop, 1935) it was found that a semantic color word that conflicted with the color the word was printed in (e.g., the word blue printed in red ink) interfered with naming of the ink color. Stroop and others more recently (e.g. Besner, Stolz, and Boutilier, 1997; Cohen, Dunbar, and McClelland, 1990) reasoned that semantic processing had a special advantage and was processed automatically even when it hindered performance. More recent investigations have challenged this automaticity explanation (Besner and Stolz, 1999; Besner et al., 1997) but the effects remain robust under the original condition. Results have consistently demonstrated that it is more difficult for subjects to respond to the color of the word than the written word itself when the two sources of information conflict and are interspersed with trials in which they do not conflict.

Barrow and Baldwin's Auditory Spatial Stroop paradigm (Barrow and Baldwin, 2010a; Barrow and Baldwin, 2010b) assessed people's ability to attend to either the spatial location or the semantics information of the stimuli. In this paradigm the words "Left" or "Right" are presented from speakers located to the left and right of the participant. In the semantic condition participants are asked to respond to the word regardless of where it is presented from. In the location condition participants are asked to respond to the location where the word is heard, regardless of which word is heard.

Similar to the results of the original Stroop (1935) experiment, Barrow and Baldwin (2009, 2010) found that it was more difficult for people to attend to the spatial location and ignore the semantics of the word, compared to the condition that instructed participants to attend to the semantics and not the spatial location of the word.

However, they also looked into individual differences based on self-reported sense of direction (SDQ; Kato and Takeuchi, 2003); these findings are of most interest. Barrow and Baldwin (submitted) found that those with a good sense-of-direction (GSD -who tend to use a visuospatial strategy) struggled the most with non-congruent spatial information which subsequently resulted in slower responses in the semantic condition. That is, those with a GSD experienced difficulty ignoring incongruent spatial information thus resulting in slower responses when instructed to respond to the semantic information. Also, those with a poor sense-of-direction experienced more difficulty when the semantic information was incongruent, resulting in slower responses when instructed to respond to the spatial location of the stimuli. Further, this pattern of results has been replicated in a recent event-related potential investigation. Results indicate that these differences as a function of navigational style and manifest differentially between the two groups of navigators as early as 200 ms after hearing the incongruent word or spatial location (Buzzell, Baldwin, Roberts, Barrow, and McDonald, 2012).

Together results from the Auditory Spatial Stroop paradigm indicate that people have fundamentally different approaches to the use of spatial information that manifest early in the processing stream at the level of orienting to a single word or sound. Individuals who rely on a verbal strategy have a strong natural inclination to process spatial information verbally and have great difficulty suppressing incongruent semantic spatial information. Conversely, visuospatial navigators have a more natural inclination to process the physical location of spatial information and find this information difficult to ignore even when it hinders task performance. Regardless of whether these individual differences are a function of innate, experiential or their combination of influences, in future investigations it would be quite interesting to determine when, developmentally, the differences emerge.

CONCLUSION

Training individuals how to utilize multiple types of spatial representations of spatial information may have far reaching implications. For example, Reagan and Baldwin (2006) found that adding landmarks to auditory route guidance systems can help people learn the environment better compared to not having landmarks. Additionally, adding cardinal direction information to an egocentric based guidance system can help build survey knowledge and maintain orientation, especially for older individuals who may not be familiar with newer route guidance systems that display navigation information from an egocentric perspective (Baldwin, 2009). Potentially the most difficult thing to get someone to learn is to use mental rotation or perspective taking strategies for orientation when he or she is severely challenged with sense of direction (Hegarty and Waller, 2004; Kozhevnikov, 2006). Because of the wide range of individual differences in navigation abilities and sense of direction, as well as preferences in navigation modality, it is important to balance a preferred method and what may be the most effective method for navigating with and the method that may facilitate route learning the best.

REFERENCES

Allen, G. L. (2004). *Human spatial memory: Remembering where*. Mahwah, NJ, US: Lawrence Erlbaum Associates Publishers.

Aretz, A. J., and Wickens, C. D. (1992). The mental rotation of map displays. *Human Performance, 5*(4), 303-328.

Baddeley, A. (2002). The concept of episodic memory. *Baddeley, Alan*.

Baddeley, A. (2003). Working Memory: Looking Back and Looking Forward. *Nature Reviews Neuroscience, 4*(10), 829-839

Baddeley, A. D. (1992). Working memory. *Science, 255*, 556-559.

Baddeley, A. D. (2000). The episodic buffer: a new component of working memory? *Trends in Cognitive Sciences, 4*(11), 417-423.

Baddeley, A. D., and Hitch, G. (1974). Working memory. In G. H. Bower (Ed.), *The psychology of learning and motivation* (Vol. 8, pp. 47-89). Orlando, FL: Academic Press.

Baddeley, A. D., and Hitch, G. J. (1994). Developments in the concept of working memory. *Neuropsychology, 8*, 4485-4493.

Baldwin, C. L. (2009). Individual differences in navigational strategy: Implications for display design. *Theoretical Issues in Ergonomics Science, 10*(5), 443-458. doi: 10.1080/14639220903106379.

Baldwin, C. L., and Reagan, I. (2009). Individual Differences in Route-Learning Strategy and Associated Working Memory Resources. *Human Factors, 51*(3), 368-377. doi: 10.1177/0018720809338187.

Barrow, J. H., and Baldwin, C. L. (2010a, September 2010). *Allowing for Individual Differences in Auditory Warning Design: Who Benefits from Spatial Auditory Alerts?* Paper presented at the Human Factors and Ergonomics Society Annual Conference San Francisco, CA.

Barrow, J. H., and Baldwin, C. L. (2010b, March 2010). *Navigation strategy matters when alerting the driver*. Paper presented at the Virginia Social Sciences Association Annual Meeting, Petersburg, VA.

Barrow, J. H., and Baldwin, C. L. (submitted). Individual Differences in Verbal-Spatial Facilitation and Conflict in Rapid Spatial Orientation.

Besner, D., and Stolz, J. A. (1999). Unconsciously controlled processing: The Stroop effect reconsidered. *Psychonomic Bulletin and Review, 6*(3), 449-455. doi: 10.3758/bf03210834

Besner, D., Stolz, J. A., and Boutilier, C. (1997). The Stroop effect and the myth of automaticity. *Psychonomic Bulletin and Review, 4*(2), 221-225. doi: 10.3758/bf03209396

Blajenkova, O., Motes, M. A., and Kozhevnikov, M. (2005). Individual differences in the representations of novel environments. *Journal of Environmental Psychology, 25*(1), 97-109.

Buzzell, G., Baldwin, C., Roberts, D., Barrow, J., and McDonald, C. (2012, March 2012). *Navigational Style and Electrophysiological Correlates of Conflict in an Auditory Spatial Stroop Task.* Paper presented at the Cognitive Neuroscience Society, Chicago, IL.

Choi, J., and Silverman, I. (2003). Processes underlying sex differences in route-learning strategies in children and adolescents. *Personality and Individual Differences, 34*(7), 1153-1166.

Cohen, J. D., Dunbar, K., and McClelland, J. L. (1990). On the control of automatic processes: A parallel distributed processing account of the Stroop effect. *Psychological Review, 97*(3), 332-361. doi: 10.1037/0033-295x.97.3.332.

Coluccia, E., Bosco, A., and Brandimonte, M. A. (2007). The role of visuo-spatial working memory in map learning: New findings from a map drawing paradigm. *Psychological Research/Psychologische Forschung, 71*(3), 359-372.

Coluccia, E., and Louse, G. (2004). Gender differences in spatial orientation: A review. *Journal of Environmental Psychology, 24*(3), 329-340.

Cooper, E. E., and Brooks, B. E. (2004). Qualitative Differences in the Representation of Spatial Relations for Different Object Classes. *Journal of Experimental Psychology: Human Perception and Performance, 30*(2), 243-256.

Cornoldi, C., and Vecchi, T. (2003). *Visuo-spatial working memory and individual differences.* New York: Psychology Press.

Frick, A., Ferrara, K., and Newcombe, N. S. (2013). Using a touch screen paradigm to assess the development of mental rotation between 3½ and 5½ years of age. *Cognitive Processing*, No Pagination Specified. doi: 10.1007/s10339-012-0534-0.

Furukawa, H., Baldwin, C. L., and Carpenter, E. M. (2004). Supporting Drivers' Area-Learning Task with Visual Geo-Centered and Auditory Ego-Centered Guidance: Interference or Improved Performance? In D. A. Vincenzi, M. Mouloua and P. A. Hancock (Eds.), *Human Performance, Situation Awareness and Automation: Current Research and Trends, HPSAA II* (pp. 124-129). Daytona Beach, FL.

Garcia, A., Finomore, V., Burnett, G., Baldwin, C. L., and Brill, C. (2012). *Individual Differences in Multimodal Waypoint Navigation.* Paper presented at the Human Factors and Erognomics Society 56th Annual Meeting, Boston, MA. .

Garden, S., Cornoldi, C., and Logie, R. H. (2002). Visuo-spatial working memory in navigation. *Applied Cognitive Psychology, 16*(1), 35-50.

Gugerty, L., and Brooks, J. (2001). Seeing where you are heading: Integrating environmental and egocentric reference frames in cardinal direction judgments. *Journal of Experimental Psychology: Applied, 7*(3), 251-266.

Gugerty, L., and Brooks, J. (2004). Reference-Frame Misalignment and Cardinal Direction Judgments: Group Differences and Strategies. *Journal of Experimental Psychology: Applied, 10*(2), 75-88.

Hegarty, M., Montello, D. R., Richardson, A. E., Ishikawa, T., and Lovelace, K. (2006). Spatial abilities at different scales: Individual differences in aptitude-test performance and spatial-layout learning. *Intelligence, 34*(2), 151-176.

Hugdahl, K., Thomsen, T., and Ersland, L. (2006). Sex differences in visuo-spatial processing: An fMRI study of mental rotation. *Neuropsychologia, 44*(9), 1575-1583.

Kato, Y., and Takeuchi, Y. (2003). Individual differences in wayfinding strategies. *Journal of Environmental Psychology, 23*(2), 171-188.

Klatzky, R. L., Loomis, J. M., Beall, A. C., Chance, S. S., & Golledge, R. G. (1998). Spatial updating of self-position and orientation during real, imagined, and virtual locomotion. *Psychological Science, 9*(4), 293-298.

Lawton, C. A. (1994). Gender differences in way-finding strategies: Relationship to spatial ability and spatial anxiety. *Sex Roles, 30*(11-12), 765-779.

Lawton, C. A. (2001). Gender and regional differences in spatial referents used in direction giving. *Sex Roles, 44*(5-6), 321-337.

Loomis, J. M., Golledge, R. G., Klatzky, R.L. (1998). Navigation system for the blind: Auditory display modes and guidance. *Presence: Teleoperators and Virtual Environments.* 7,193–203.

MacFadden, A., Elias, L., and Saucier, D. (2003). Males and females scan maps similarly, but give directions differently. *Brain and Cognition, 53*(2), 297-300.

Moffat, S. D., Zonderman, A. B., and Resnick, S. M. (2001). Age differences in spatial memory in a virtual environment navigation task. *Neurobiology of Aging, 22*(5), 787-796.

O'Keefe, J., and Nadel, L. (1978). *The Hippocampus as a Cognitive Map.* Oxford, England: Oxford University Press.

Pazzaglia, F., and De Beni, R. (2001). Strategies of processing spatial information in survey and landmark-centred individuals. *European Journal of Cognitive Psychology, 13*(4), 493-508.

Reagan, I., and Baldwin, C. L. (2006). Facilitating route memory with auditory route guidance systems. *Journal of Environmental Psychology, 26*(2), 146-155.

Ribordy, F., Jabès, A., Banta Lavenex, P., and Lavenex, P. (2013). Development of allocentric spatial memory abilities in children from 18 months to 5 years of age. *Cognitive Psychology, 66*(1), 1-29. doi: 10.1016/j.cogpsych.2012.08.001.

Richardson, A., Montello, D., Hegarty, M., and (1999). Spatial knowledge acquisition from maps and from navigation in real and virtual environments. *Memory and Cognition, 27*(4), 741-750.

Saucier, D., Bowman, M., and Elias, L. (2003). Sex differences in the effect of articulatory or spatial dual-task interference during navigation. *Brain and Cognition, 53*(2), 346-350.

Streeter, L. A., Vitello, D., and Wonsiewicz, S. A. (1985). How to tell people where to go: Comparing navigational aids. *International Journal of Man Machine Studies, 22*(5), 549-562.

Stroop, J. R. (1935). Studies of interference in serial verbal reactions. *Journal of Experimental Psychology, 28*, 643-662.

Thorndyke, P. W., and Hayes-Roth, B. (1982). Differences in spatial knowledge acquired from maps and navigation. *Cognitive Psychology, 14*(4), 560-589.

Tolman, E. C. (1924). The Inheritance of Maze-Learning Ability in Rats. *Journal of Comparative Psychology, 4*(1), 1-18.

Tolman, E. C. (1948). Cognitive maps in rats and men. *Psychological Review, 55*(4), 189-208.

Van Erp, J. B. F., and Van Veen, H. (2004). Vibrotactile in-vehicle navigation system. *Transportation Research Part F-Traffic Psychology and Behaviour, 7*(4-5), 247-256. doi: 10.1016/j.trf.2004.09.003.

Van Erp, J. B. F., Van Veen, H. A. H. C., and Jansen, C. (2006). Waypoint navigation with a vibro-tactile waist belt. *ACM transactions on applied perception, 2*(2), 106-117.

Voyer, D., Voyer, S., and Bryden, M. P. (1995). Magnitude of sex differences in spatial abilities: A meta-analysis and consideration of critical variables. *Psychological Bulletin, 117*(2), 250-270. doi: 10.1037/0033-2909.117.2.250.

Chapter 10

WORKING MEMORY IMPROVEMENT BY THE DIFFERENTIAL OUTCOMES PROCEDURE

Ginesa López-Crespo[1] and Angeles F. Estévez[2]
[1]Department of Psychology and Sociology. University of Zaragoza, Zaragoza, Spain
[2]Department of Psychology. University of Almería, Almería, Spain

ABSTRACT

A growing body of evidence shows that the arrangement of differential outcomes (DO) can increase the retention of stimuli on several working memory (WM) tasks. The differential outcomes effect (DOE) was initially described in the field of animal learning by Trapold (1970). Trapold showed that when each of the different correct sample-choice sequences of a conditional discrimination problem is followed by a unique outcome, learning is faster and the performance is better than when all correct responses are followed by a common outcome (the non-DO condition). Subsequently, findings from several animal studies suggest that this procedure can also improve memory performance (i.e., Brodigan and Petersen, 1976). Recently, research has shown the effectiveness of DO in memory tasks such as delayed matching to position in a variety of species (i.e., rats and pigeons), and also in human beings (i.e., Hochhalter, Sweeney, Bakke, Holub, and Overmier, 2000; López-Crespo, Daza, and Méndez-López, 2011; López-Crespo, Plaza, Fuentes, and Estévez, 2009; Plaza, Estévez, López-Crespo, and Fuentes, 2011; Plaza, López-Crespo, Antúnez, Fuentes, and Estévez, 2012). This chapter reviews and discusses the results of these studies in terms of their theoretical considerations, biological underpinnings, and applied potential of the effect.

INTRODUCTION

Sometimes, a small change in something we do routinely in our daily life can lead to great improvements. Imagine that we are trying to teach young children a discrimination problem – to stop when the traffic light is red and to cross the street if the light is green. A usual way to encourage this sort of discriminative learning is to provide any reward after each correct response –for example saying to the child "well done". But what would happen if we instead provide the child with a specific outcome for each correct response –for instance, a kiss when he(she) correctly chooses to cross the street and the phrase 'well done' when he(she) correctly chooses to stop? Trapold (1970) answered this question in a conditional discrimination study with animals. He exposed rats to a discrimination problem that required a response to one lever (e.g., the right lever) in the presence of one stimulus (e.g., a tone), and a different response to a second lever (e.g., the left lever) in the presence of another stimulus (e.g., a click). Trapold observed increased rate of acquisition and greater accuracy when the correct choice of the right lever was followed by pellets and the correct choice of the left lever was followed by sucrose, than when both correct responses produced the same reinforcer. This enhancement in performance and/or terminal accuracy was termed the differential outcomes effect (DOE; Trapold and Overmier, 1972).

The improvement of conditional discrimination learning by differential outcomes appears to be a ubiquitous phenomenon. Its effectiveness has been demonstrated in a variety of species, including pigeons, rats, dogs (see Goeters, Blakely, and Poling, 1992, for a review) and human beings. The results found in humans show that the differential outcomes procedure (DOP) is effective in improving the learning of symbolic relations in children ranged in age from four years to eight years and six months (Estévez and Fuentes, 2003; Estévez, Fuentes, Marí-Beffa, González, and Álvarez, 2001; Maki, Overmier, Delos, and Gutman, 1995) as well as in adults without mental handicaps (Estévez, Vivas, Alonso, Marí-Beffa, Fuentes, and Overmier, 2007; Miller, Waugh, and Chambers, 2002; Mok and Overmier, 2007).

Moreover, the benefits of using the DOP have also been extended to different clinical populations such as children with mental retardation (Jansen and Guess, 1978; Saunders and Sailor, 1979), children with autism (Litt and Schreibman, 1981), adults with Prader-Willy syndrome (Joseph, Overmier, and Thompson et al., 1997) and children and adults with Down's syndrome (Estévez, Fuentes, Overmier, and González, 2003).

Importantly, the DOP has also been shown to improve memory- based performance after delays in animals. For example, Brodigan and Peterson (1976) used a delayed matching-to-sample (DMTS) task with pigeons, a procedure that traditionally has served to assess working memory in basic research with laboratory animals (Rodríguez and Paule, 2009). Each trial began with the presentation of a sample stimulus, a red or a green light. After a variable delay (0, 3, or 15 seconds), subjects were required to choose between two comparison stimuli, a horizontal or a vertical line presented on two response levers. When the sample was the red light, the vertical line was the correct choice, and when the sample was the green light, the correct choice was the horizontal line. Results showed that in the non-differential condition performance lowered to chance after a delay of a few seconds between the sample and the comparison stimuli; however, the application of differential consequences improved the participants' performance even at the longer delays.

The experiment of Brodigan and Petersen (1976) was the first demonstration of a positive effect of the DOP on short-term memory. Just like on discrimination learning, this memory improvement has been demonstrated in a variety of species including pigeons, rats, and human beings (see also Goeters et al., 1992 for a review) as well as in a variety of task and conditions, as developed in the next section.

A UBIQUITOUS PHENOMENON

In addition to the DMTS procedure employed by Brodigan and Petersen (1976), other tasks have been used to study the DOE. For example, delayed matching to position (DMP) tasks are frequently used in animals. In an operant version of this task, a lever is inserted on an operant chamber and retracted once the animal makes a lever press. After a variable delay, the two levers are presented and the animal is required to press the sample lever to obtain the reinforcer (Savage and Parsons, 1997, Savage, Pitkin, and Careri, 1999). Human experiments have used mainly DMTS tasks with different sample and choice stimuli (for example, human faces, geometric shapes or visual scenes) and outcomes (see below). Interestingly, the DOE benefits extend also to spatial memory tasks. In a recent study, Legge and Specht (2009) explored the effect of the DOP in a group of university students performing a landmark-based spatial memory task. In one of the experiments, the task involved the presentation of a sample object that was associated with a particular landmark-to-goal relationship. The results showed that search accuracy was increased when differential outcomes were arranged. Similarly, we have also obtained in our laboratory (in preparation) that five- and seven-year-old children showed better spatial memory recognition when each location to be remembered was paired with its own outcome.

Another source of variation in the DOE experiments is the control condition. There are two usual ways to arrange the control, non-differential condition. In one of them, subjects always receive the same reinforcer or outcome (for example, in the differential condition correct responses to one sample are followed by a food pellet and correct responses to the other sample are paired with sucrose, whereas in the control condition all correct responses are always paired with a food pellet). This arrangement is usually termed "common" condition. In other experiments, the outcomes are randomly arranged (for example, in the control condition a food pellet and sucrose is programmed to appear randomly after correct responses). This procedure is termed random outcomes or non-differential outcomes (non-DO) procedure or condition (Mok, Thomas, Lungu, and Overmier, 2009; Savage, 2001).

Even the outcome itself differs between experiments. It is important to note that not only rewarding events such as food or water can give rise to a DOE in memory tasks. DOE has been showed in animals with different sensory outcomes (blue vs. yellow lights; Kelly and Grant, 2001) or no food versus food delivery (Urcuioli, 1990). Moreover, not only different types of reinforcers serve as differential outcomes (for example, sucrose vs. water) but different probabilities of reinforcement do. For example, in the study of DeLong and Wasserman (1981) response to one correct stimulus produced reinforcement (food) with a probability of 0.2, whereas responses to the other stimulus produced the same reinforcer with a probability of 1.0. In the control condition, correct responses to both stimuli produced reinforcement with a probability of 0.6. The result showed that acquisition was faster and terminal accuracy greater in the DOE condition than in the control condition. Therefore, the motivational properties of the outcomes are not the key factor to produce DOE. Effective outcomes used with humans includes photographs along with primary reinforcers (i.e., an umbrella or a mug) that were raffled at the end of the study (i.e., López-Crespo et al., 2009; Plaza et al., 2011), different perceptual outcomes (i.e., visual vs. auditory; e.g., Mok et al., 2009) or even different stimuli in the same modality (i.e., López-Crespo et al., 2012).

In spite of these and other sources of variation in DOE studies not mentioned here (for example, the DOE is evident when both enter and intra subjects designs are used; e.g., Hochhalter et al., 2000; López-Crespo et al., 2009; Martella, Plaza, Estévez, Castillo, and Fuentes, 2012; Mok et al., 2009; Plaza et al., 2012), the results are quite consistent. Namely, a common result of most of these studies is that the arrangement of differential outcomes enhances discriminative learning and/or recognition memory in animals and in human beings. Regarding the value of the DOP to short-term memory, Petersen and Trapold (1980) state that "with differential outcomes, subjects are able to maintain very good performance in spite of substantial delays".

THEORETICAL ACCOUNTS OF THE DOE

Early learning psychologists such as Thorndike (1911) and Hull (1943) depicted discrimination learning as the development of a stimulus-response association (S-R association). The reinforcing outcome (O) was mostly considered to be just a catalyst for this association, but was not itself part of what was learned serving only to strengthen the association between the stimulus and the response. Thus, according to this perspective, discriminative learning would depend critically on whether the S-R sequences are followed by rewards but not on how they are arranged.

The development of the two-process theory of learning (Mowrer, 1947, 1956) acknowledged the role of outcomes on associative learning. According to this theory two associations are learned in discrimination learning: an S-R association plus an S-O association. On the basis of this theory, Trapold (1970) demonstrated that discrimination learning proceeds more rapidly when different responses produce different outcomes (i.e., reinforcers) than when they produce the same one. In his seminal paper Trapold (1970) hypothesized that the DOE occurs because when each sequence of stimulus (i.e. to respond to the left lever in presence of a tone and to respond to the right lever in presence of a light) was followed by an unique outcome event (for example, a food pellet and sucrose, respectively), the subject learnt an expectancy of this particular outcome, and this expectancy provided an additional source of information that guided the choice behaviour. That is, the expectancy acquires stimulus control over the instrumental performance. Although subjects that receive the same outcome for different correct S-R sequences develop also outcome expectancies, because the expectancies are common to both choices, they can contribute nothing to enhance accuracy.

As mentioned earlier, the arrangement of differential outcomes improves memory performance even at longer delays. It seems that the expectancies could bridge the time delay between the removal of a discriminative stimulus and the response to the choice stimulus (Overmier and Linwick, 2001). But, by which mechanisms does the DOP produce such enhanced memory performance? Recently, in an interesting extension of the aforementioned expectancy theory proposed by Trapold (1970, see also Overmier and Trapold, 1972), Savage and colleagues (v.g., Overmier, Savage, and Sweeney, 1999; Savage, 2001; Savage et al., 1999) proposed that under non-differential outcomes conditions, the only source of information that can guide the correct response is a explicit retrospective memory process that maintains the discriminative stimulus over the delay. However, the DOP result in the formation of a unique reward expectancy (or prospective memory) that is learned via S-O associations, that is, via pavlovian conditioning. And classical conditioning is a kind of unintentional learning representative of implicit memory process.

If the DO and non-DO procedures are indeed mediated by two different memory systems as Savage and colleagues have proposed, researchers must be able to find double dissociations between them. Studies aiming to explore the biological underpinnings of DOE have found evidences of these double dissociations.

BIOLOGICAL UNDERPINNINGS OF THE DOE

Using a pharmacological approach, Savage and Parsons (1997) conducted a study in which a double dissociation between differential and non-differential conditions was first demonstrated. These authors administered different doses of the cholinergic antagonist scopolamine and the NMDA glutamatergic receptor antagonist MK-801 to rats performing a DMTP task under differential and non-differential conditions. Results showed that scopolamine disrupted performance under non-differential, but not under differential conditions. The reverse results were found for MK-801: this drug disrupted performance under differential, but not under non-differential conditions.

In a more recent study of the same group, Savage, Buzzeti, and Ramirez (2004) demonstrated that performance of rats with hippocampal lesions was dramatically impaired when they were trained under non differential conditions, whereas the same lesion had only a transient effect that disappeared by the end of training in subjects trained under differential conditions. Thus, it appears that the hippocampus is crucial for performance under non-differential conditions. By contrast, the basolateral amygdala (BLA) appears to be the structure underlying performance under differential conditions, as showed by Savage, Koch, and Ramirez (2007). In their study, BLA inactivation with microinjections of the γ-aminobutyric acid (GABA) agonist muscimol, dose-dependently impaired performance of rats trained under differential outcomes, but it had no effect on animals trained under non-DO.

These data are consistent with the two-memory systems model of the DOE since hippocampal damage has been associated with impairments of declarative/explicit or episodic aspects of a task, but not with implicit aspects (Cohen and Squire, 1980, Schacter, 1998; Squire, 1992). By contrast, several studies have shown that reward-related learning relies on different structures as, for example, the amygdala (Savage and Ramos, 2009).

An interesting study conducted by Mok et al. (2009) also supports the two memory system model by showing that the brain areas recruited under DO and non-DO conditions are different in human beings. In this study participants had to perform a symbolic discrimination task. The two sample stimuli and the four choice alternatives were geometrical figures separated by a 3.5 sec delay. For one group of participants, correct choices to one sample stimulus were followed by a visual outcome (pictures of smiling babies) and correct choices to the other sample stimulus were followed by an auditory outcome (the song Macarena) (differential condition, DO). The same group conducted a similar, equated in difficulty, task, but in this case the outcomes after correct responses to both cues were identical (a combined visual and auditory outcome) (common condition, CO). Another group of participants performed an identical task with the difference that the outcomes were randomly delivered (non-differential condition, NDO). Overall, participants performed the task significantly better in the DO condition than in the CO condition. fMRI data were recorded on the delay period. Results showed greater hippocampal activation when non-differential outcomes were arranged suggesting that this region plays a role in mediating retrospective rather than prospective memory. Interestingly, in the DO condition, brain regions related to sensory-specific cortices were activated as a function of the outcomes being used. That is, visual association and primary visual cortices were activated, including fusiform areas involved in face-processing, when visual outcomes (pictures of smiling babies) were employed. By contrast, the primary and association auditory cortices increased delay-period activity for auditory outcomes condition. Therefore, it appears that, as Savage and colleagues hypothesized, some kind of prospective memory (a perceptual representation of the reinforcer) is activated when the sample stimulus is presented. It is also interesting to note that modality independent activation of the angular gyrus of the posterior parietal cortex was also observed in the study.

Although some findings from animal and humans studies about the brain areas involved in the DOE (amygdala vs. sensory-specific cortices) appear to be contradictory, this is not necessarily the case. An obvious difference between the two set of studies is the type of outcome being used. In the animal studies, the outcomes were rewards with an obvious hedonic value so that areas related with the brain reward system, as the BLA, were activated in the DO condition. By contrast, the outcomes used in the study of Mok et al. (2009) were sensory-perceptual events, and therefore, sensory areas were recruited under differential outcomes.

APPLICABILITY OF THE DOP IN CLINICAL POPULATIONS

Animal studies provided early clues about the applicability of the DOP on memory deficits. Indeed, recent animal studies have demonstrated that this procedure can reverse working memory deficits in animal models of the Wernicke-Korsakoff syndrome (Savage and Langlais, 1995) and in aged rats (Savage et al., 1999). In the first study, rats were treated with pyrithiamine until toxicity that caused central nervous system lesions that paralleled those seen in patients with Korsakoff's syndrome and, in turn, performance impairments on tasks that assessed working memory (i.e., DMTP). Savage and Langlais (1995) observed an enhancement of both acquisition and DMTP performance of rats treated with pyrithiamine receiving differential outcomes. In fact, their memory performance was comparable to that of normal control rats. Using a similar procedure, Savage et al. (1999) also found that the DOP enhanced memory performance in aged rats on a DMTP task for which they normally were impaired. Moreover, these rats did not display the typical age-related decline in spatial working memory when differential outcomes were arranged.

The results from these studies led some researchers to explore whether similar benefits of the DOP would be obtained in human populations with memory impairments. In an extension of the work reported by Savage and Langlais (1995), Hochhalter et al. (2000) trained four patients with alcohol dementia to recognize which of two faces matched a previously seen face, a task that these patients found difficult to solve. One patient did not show any difference in matching accuracy when trained with differential and non-differential outcomes, and his data were not included in the statistical analyses. The other three patients with memory impairments performed the task more accurately when differential outcomes were arranged although the effect was mainly evident at the five-second delay. In fact, their performance did not differ from that of controls at that delay. At longer delays, however, patients showed a low accuracy regardless of the type of training used.

To test whether similar results as those obtained with aged rats (Savage et al., 1999) could be observed in aged people, López-Crespo et al. (2011) conducted a study comparing the performance of old and young people in a DMTS task. Participants were exposed to a man's face, and after a delay, they were required to decide if the previously seen face was within a set of six men's faces. For half of the subjects, each sample face was paired with its own outcome (differential outcomes condition); outcomes were randomly arranged for the remaining half of subjects (non-differential condition). Either short (5 sec) or long (30 sec) delays were interposed between the sample and the comparison stimuli. Results showed that relative to younger adults, older adults' performance decreased with the longer delay. However, the arrangement of differential outcomes was able to reverse the detrimental effect of the increased delay usually observed in the elderly group, raising their performance up to the level shown by younger adults.

Similar results were found by Plaza et al. (2012) with Alzheimer's disease(AD) patients. Eight patients with AD were tested using a within-subject DMTS task similar to that employed by López-Crespo et al. (2011). The results showed that these patients were less accurate and their reaction times (RT) were lower when training under the non-DO condition. However, when DO where arranged, they showed a significantly better and faster delayed face recognition even at the longer delay (25sec).

López-Crespo et al. (2012) examined the DOE in deaf children with diverse communication modes: spoken Spanish (oral), Spanish Sing Language (SSL) and both spoken Spanish and SSL (that is, bilingual). A within-subjects, DMTS task was employed. The task included visual, not easily verbalizable, complex stimuli (Kanji letters) to specifically test the visual WM of the children. Briefly, a kanji sample letter appeared in the centre of the screen, followed by a delay of 0 or 4 seconds. Then a comparison stimulus appeared; 50% of times it was identical to the sample stimulus whereas in the remaining 50% it differed on a small attribute. SSL and oral children were less accurate than bilingual and control children on the DMTS task. Analysis on reaction times showed higher RTs at the longer delay (4 sec) for all the deaf groups (SSL, oral, and bilinguals) than for control children. The use of the DOP increased accuracy and lowered RTs in all the groups tested.

The benefits of using the DOP have also been demonstrated in children born prematurely. Martinez et al. (2012) conducted a study using a delayed visuospatial recognition task that consisted of the presentation of a two-dimensional geometric sample figure that was replaced, after a delay of 2 seconds, by four comparison figures. One of these figures, the correct choice, was the sample figure rotated clockwise by 90, 180 or 240 degrees. The results revealed that the percentage of correct choices was higher for control than for the premature seven-year-old children under non-differential conditions. This difference disappeared when children received specific outcomes following their correct responses. That is, visuospatial recognition of children born prematurely was improved by using differential outcomes.

Finally, Martella et al. (2012) have shown that the DOP can ameliorate the working memory deficits observed in normal adults suffering from a transient memory problem due to sleep deprivation. Participants had to perform a delayed face recognition task similar to that used by Plaza et al. (2011; Experiment 2b). On each trial six faces (sample stimuli) appeared for 4 sec. After a variable delay of 5, 10, 25, or 32 sec, the comparison face was presented. They had to decide whether this face was present among the six sample faces. The results showed that accuracy was lower in sleep-deprived subjects than in non-deprived controls. In fact, their percentage of correct responses was at chance level for all delays. Interestingly, the DOP improved the recognition performance of these participants, mainly with short delays.

CONCLUSION

The studies reviewed in this chapter show that the DOE is a ubiquitous phenomenon that has been demonstrated in a variety of conditions and populations, including children, adults and older adults with and without memory problems. Taken together, the results from these studies indicate the potential of the differential outcomes training as a therapeutic technique to assist those people with short-term memory deficits. Future basic research should undoubtedly investigate which are the ideal conditions and populations that might be benefited by the DOP and the impact that this might, in turn, have in their daily life. Then, it will be the moment to use this procedure in more applied settings, such as instructional ones. To provide an example, we are immersed in a high-tech world where almost all parents in developed countries have a considerable amount of mobile applications. Why not create applications that help our children to easily learn and remember simple discriminations such as those between colours and letters through differential outcomes?

REFERENCES

Brodigan, D.L., and Peterson, G.B. (1976). Two-choice conditional discrimination performance of pigeons as a function of reward expectancy, prechoice delay and domesticity.*Animal Learning and Behavior, 4*, 121–124.

Delong, R.E., and Wasserman, E.A. (1981). Effects of differential reinforcement expectancies on successive matching-to-sample performance in pigeons. *Journal of Experimental Psychology: Animal Behavior Processes, 7,* 394-412.

Cohen, N.J., and Squire, L.R. (1980). Preserved learning and retention of pattern analyzing skill in amnesia: Dissociation knowing how and knowing what. *Science, 210,* 207-209.

Estévez, A.F., and Fuentes, L.J. (2003). Differential outcomes effect in four-year-old children. *Psicológica, 24,* 159–167.

Estévez, A.F., Fuentes, L.J., Marí-Beffa, P., González, C., and Alvarez, D. (2001). The differential outcomes effect as useful tool to improve conditional discrimination learning in children. *Learning and Motivation, 32,* 48–64.

Estévez, A.F., Overmier, J.B., Fuentes, L.J., and González, C. (2003). Differential outcomes effect in children and adults with Down syndrome. *American Journal on Mental Retardation, 108,* 108–116.

Estévez, A.F., Vivas, A.B., Alonso, D., Marí-Beffa, P., Fuentes, L.J., and Overmier, J.B. (2007). Enhancing challenged students' recognition of mathematical relations through differential outcomes training. *Quarterly Journal of Experimental Psychology, 60,* 571 – 580.

Goeters, S., Blakely, E., and Poling, A. (1992). The differential outcomes effect. *The Psychological Record, 42,* 389–411.

Hochhalter, A.K., Sweeney, W.A., Bakke, B.L., Holub, R.J., and Overmier, J.B. (2000). Improving face recognition in alcohol dementia. *Clinical Gerontologist, 22(2),* 3-18.

Hull, C.L. (1943). *Principles of behavior: An introduction to behavior theory.* New York: Appleton-Century-Crofts.

Janssen, C., and Guess, D. (1978). Use of function as a consequence in training receptive labeling to severely and profoundly retarded individuals. *AAESPH Review, 3,* 246-258.

Joseph, B., Overmier, J.B., and Thompson, T. (1997). Food and nonfood related differential outcomes in equivalence learning by adults with Prader – Willi syndrome. *American Journal on Mental Retardation, 4,* 374–386.

Kelly, R., and Grant, D.S. (2001). A differential outcomes effect using biologically neutral outcomes in delayed matching-to-sample with pigeons. *The Quarterly Journal of Experimental Psychology, 54B,* 69–79.

Legge, E.L., and Spetch, M.L. (2009). The differential outcomes effect (DOE) in spatial localization: An investigation with adults. *Learning and Motivation, 40,* 313–328.

Litt, M.D., and Schreibman, L. (1981). Stimulus- specific reinforcement in the acquisition of receptive labels by autistic children. *Analysis and Intervention in Developmental Disabilities, 1,* 171–186.

López-Crespo, G., Daza, M.T., and Méndez-López, M. (2012). Visual working memory in deaf children with diverse communication modes: improvement by differential outcomes. *Research in Developmental Disabilities, 33,* 362-368.

López-Crespo, G., Plaza, V., Fuentes, L.J., and Estévez, A.F. (2009). Improvement of age-related memory deficits by differential outcomes. *International Psychogeriatrics, 21,* 503-510.

Martella, D., Plaza, V., Estévez, A.F., Castillo, A., and Fuentes, L.J. (2012). Minimizing sleep deprivation effect in healthy adults by differential outcomes. *Acta Psychologica, 139,* 391-396.

Maki, P., Overmier, J.B., Delos, S., and Gutman, A.J. (1995). Expectancies as factors influencing con- ditional discrimination performance of children. *The Psychological Record, 45,* 45–71.

Martínez, L., Marí-Beffa, P., Roldán-Tapia, D., Ramos-Lizana, J., Fuentes, L.J., and Estévez, A.F. (2012). Training with differential outcomes enhances discriminative learning and visuospatial recognition memory in children born prematurely. *Research in Developmental Disabilities, 33,* 76-84.

Miller, O.T., Waugh, K.M., and Chambers, K. (2002). Differential outcomes effect: Increased accuracy in adults learning kanji with stimulus specific rewards. *The Psychological Record, 52,* 315–324.

Mok, L.W., and Overmier, J.B. (2007). The differential outcomes effect in normal human adults using a concurrent-task within-subjects design and sensory outcomes. *The Psychological Record, 57,* 187 – 200.

Mok, L.W., Thomas, K.M., Lungu, O.V., and Overmier, J.B. (2009). Neural correlates of cue-unique outcome expectations under differential outcomes training: An fMRI study. *Brain Research, 1265,* 111–127.

Mowrer, O.H. (1947). On the dual nature of learning- a reinterpretation of "conditioning" and "problem solving". *Harvard Educational Review, 17*, 02–148.

Mowrer, O.H. (1956). Two-factor learning theory reconsidered, with special refer- ence to secondary reinforcement and the concept of habit. *Psychological Review, 63*, 114–28.

Overmier, J.B.,and Linwick, D. (2001). Conditional choice-unique outcomes establish expectancies that mediate choice behavior. *Integrative Psysiological and Behavioral Science, 36(3)*, 173-181.

Overmier, J.B., Savage, L.M., and Sweeney, W.A. (1999). Behavioral and pharmacological analyses of memory: New behavioral options for remediation. In M. Haugand R.E. Whalen (Eds.), *Animal models of human emotion and cognition* (pp. 231– 245). Washington, DC: American Psychological Association.

Peterson, G.B., and Trapold, M.A. (1980). Effects of altering outcome expectancies on pigeons' delayed conditional discrimination performance. *Learning and Motivation, 11*, 267-288.

Plaza, V., López-Crespo, G., Fuentes, L.J., and Estévez, A.F. (2011). Enhancing recognition memory in adults through differential outcomes. *Acta Psychologica, 136*, 129-136.

Plaza, V., López-Crespo, G., Antúnez, C., Fuentes, L.J., and Estévez, A.F. (2012). Improving delayed face recognition in Alzheimer's disease by differential outcomes. *Neuropsychology, 26*, 483-489.

Rodriguez, J.S., and Paule, M.G. (2009). Working Memory Delayed Response Tasks in Monkeys. In J.J. Buccafusco (Ed.), *Methods of Behavior Analysis in Neuroscience*, Chapter 12, 2nd edition. Boca Raton (FL): CRC Press.

Savage, L.M. (2001). In search of the neurobiological underpinnings of the differential outcomes effect. *Integrative Physiological andBehavioral Science, 36*, 182 – 195.

Savage, L.M., Buzzetti, R.A., and Ramirez, D.R., 2004. The effects of hippocampal lesions on learning, memory, and reward expectancies. *Neurobiology of Learning and Memory, 82*, 109–119.

Savage, L.M., Koch A.D., and Ramirez, D.R. (2007). Basolateral amygdala inactivation by muscimol, but not ERK/MAPK inhibition, impairs the use of reward expectancies during working memory. *European Journal of Neuroscience, 26*, 3645-51.

Savage, L.M., and Langlais, P.J. (1995). Differential outcomes attenuate memory impairments on matching-to-position following pyrithiamine- induced thiamine deficiency in rats. *Psychobiology, 23*, 153–160.

Savage, L.M., and Parsons, J.P. (1997). The effects of delay-interval, inter-trial interval, amnestic drugs, and differential outcomes on match- ing to position in rats. *Psychobiology, 25*, 303–312.

Savage, L.M., Pitkin, S.R., and Careri, J.M. (1999). Memory enhancement in aged rats: The differential outcomes effect. *Developmental Psychobiology, 35*, 318–327.

Savage, L.M., and Ramos, R.L. (2009). Reward expectation alters learning and memory: The impact of the amygdala on appetitive-driven behaviors. *Behavioural Brain Research, 198*, 1-12.

Schacter, D.L. (1994). Priming and multiple memory systems: Perceptual mechanisms of implicit memory. In D.L. Schacter and E. Tulving (Eds.), *Memory systems 1994* (pp. 233-268). Cambrigde, M.A.: MIT Press.

Saunders, R., and Sailor, W. (1979). A comparison of three strategies of reinforcement on two- choice learning problems with severely retarded children. *AAESPH Review, 4*, 323–333.

Squire, L.R. (1992). Memory and the hippocampus: A synthesis form finding with rats, monkeys, and humans. *Psychological Review, 99*, 195-231.

Thorndike, E.L. (1911). *Animal intelligence.* New York: Macmillan.

Trapold, M.A. (1970). Are expectancies based upon different positive reinforcing events discriminably different? *Learning and Motivation, 1*, 129-140.

Trapold, M.A., andOvermier, J.B., 1972. The second learning process in instrumental learning. In: Black, A.H., Prokasy, W.F. (Eds.), *Classical Conditioning II: Current Research and Theory* (pp. 427–452). Appleton-Century-Crofts, New York.

Urcuioli, P.J. (1990). Some relationships between outcome expectancies and sample stimuli in pigeons' delayed matching. *Animal Learning and Behavior, 18*, 302–314.

In: Working Memory
Editor: Helen St. Clair-Thompson

ISBN: 978-1-62618-927-0
© 2013 Nova Science Publishers, Inc.

Chapter 11

BILATERAL TRANSCRANIAL ALTERNATING CURRENT STIMULATION (tACS) OF THE DORSOLATERAL PREFRONTAL CORTEX ENHANCES VERBAL WORKING MEMORY AND PROMOTES EPISODIC MEMORY AFTER-EFFECTS

Oded Meiron[1] and Michal Lavidor[1,2]
[1]The Gonda Multidisciplinary Brain Research Center,
Bar Ilan University, Ramat Gan, Israel
[2]Department of Psychology, University of Hull, Hull, UK

ABSTRACT

The present study investigated transcranial alternating current stimulation (tACS) effects on working memory (WM) maintenance and memory consolidation in healthy human participants. We examined the behavioral effects of theta-frequency stimulation to the dorsolateral prefrontal cortex (DLPFC) as recent electrophysiological findings indicated that increased DLPFC theta synchronization predicts WM function and successful verbal-memory encoding. We utilized a verbal *n*-Back task known to be associated with DLPFC function and frontal theta activity to assess tACS manipulation effects on *online* (stimulation during WM task) WM accuracy, as well as on *post-stimulation* (immediately after stimulation) WM accuracy. Additionally, to investigate possible after-effects of tACS 20 minutes post-stimulation, we administered a free-recall procedure to evaluate episodic retrieval accuracy as an indication of successful memory consolidation. Results indicated enhanced online WM accuracy in the active bilateral DLPFC tACS condition. Significant *implicit* episodic memory after-effects were found in the active left DLPFC tACS condition as well as in the active bilateral DLPFC tACS condition. Most interestingly, *explicit* episodic retrieval was enhanced only in the active bilateral DLPFC condition. Our findings imply that bilateral DLPFC oscillatory stimulation may be used to increase functional connectivity in order to improve WM function and episodic memory formation.

INTRODUCTION

Successful memory maintenance and consolidation are associated with long-term plastic changes in limbic structures (anterior cingulated cortex and hippocampus, respectively) and interconnected prefrontal areas (Buzsaki 1996; Chrobak and Buzsaki 1996; Caplan et al., 2003; Cooke and Bliss, 2006; Griesmayr et al., 2010; Paz, Bauer and Pare, 2011). Additionally, electrophysiological findings indicate that the extent of "on-line" theta (4-8Hz) synchronization is related to verbal encoding, as well as predictive of successful word-recognition in humans (Klimesch, 1999). Importantly, increased power of cortical theta-band synchronization patterns are associated with sustained and significantly enhanced synchronized synaptic potentials (Caplan et al., 2003). Caplan et al. (2003) suggested that human cortical theta rhythm serves to coordinate or "synchronize" activity between forebrain limbic structures and other neocortical areas (e.g., entorhinal cortex, prefrontal cortex), or between pairs of cortical areas during exploratory learning behaviors. In regard to working memory (WM) maintenance and theta activity, Wu et al.'s (2007) EEG findings revealed greater theta power and significant large–scale theta phase synchronization between bilateral frontal regions when subjects were maintaining bound verbal-location targets in WM in comparison to maintaining verbal and spatial targets separately. In another EEG study that used a standard WM task with digit-words as studied stimuli (Schack et al., 2002), increased power in theta and gamma bands was observed during the retention interval (1800ms) in the left frontal region below electrodes Fz and Fpl, respectively, an area that includes the left dorsolateral prefrontal cortex (DLPFC). Animal models of memory consolidation have indicated enhanced neuronal-network synchronization originating in phase-related discharge from the hippocampus back to neocortical structures after an initial learning phase (Chrobak and Buzsaki, 1996). Chronbak and Buzsaki (1996) suggested that the neurophysiological role of organized network bursts could be conceptualized as long-term alterations in synaptic efficacy. Rutishauser et al. (2010) examined the relationship between spike field coherence (SFC) during theta oscillations (generated in the hippocampus) and later memory retrieval (using implanted depth electrodes with epileptic patients) and found that theta-range (3-7Hz) SFC was 50% higher in trials that were later remembered versus trials that were later forgotten. In humans, deeper verbal encoding, which subsequently leads to better memory consolidation (i.e. enhanced episodic retrieval) has been shown to be associated with enhanced left frontal activity versus right frontal activity (Tulving et al., 1994). Positron emission tomography (PET) studies that examined verbal encoding effects on explicit memory performance, found significantly higher levels of blood flow in the left prefrontal cortex in comparison to its contralateral regions. In correspondence, in PET studies that measured regional cerebral blood flow during episodic retrieval, right prefrontal activation was most prominent in comparison to left prefrontal activation (Tulving et al. 1994). As a result of consistent hemispheric differences between verbal memory-encoding and memory-retrieval observed in PET studies, as well as Rutishauser et al.'s findings (2010) related to hippocampal theta synchronization during memory encoding, we adopt the theoretical standpoint that increased left prefrontal theta activity is generated by the hippocampus during episodic verbal encoding. Alternatively, during episodic retrieval there is a higher level of hippocampus-generated theta synchronization with the right prefrontal cortex. Thus, since WM tasks require immediate encoding and retrieval within a short period, we hypothesize that continuous "online" (during WM activity) bilateral prefrontal theta-stimulation is likely to benefit WM performance, while left prefrontal theta-stimulation may enhance episodic verbal encoding resulting in enhanced episodic retrieval. Critical to our investigation, Griesmayr et al. (2010) suggested that frontal theta activity (4-6Hz) reflects increased demands on WM maintenance of studied material. Importantly, they indicated that this particular regulatory WM operation (i.e., active WM maintenance, see Kane and Engle, 2002) is controlled by the prefrontal cortex. Their findings are in agreement with our theoretical premise that prefrontal theta oscillatory activity and regulatory WM functions are co-dependent. Hence, in contrast to non-manipulative WM storage (i.e., non-active WM maintenance), simultaneous storage and manipulation are required to elicit the monitoring component of WM function (see *central executive* function, Baddeley, 1986), subsequently increasing demands on DLPFC involvement (Kane and Engle, 2002), and generating increased prefrontal theta-band synchronization patterns associated with WM performance (Wu et al., 2007; Griesmayr et al., 2010). In accordance with the "monitoring" role of WM theta oscillations, we intend to measure theta-stimulation effects on the executive attention function of WM (Kane and Engle, 2002) by using a highly demanding, prefrontal verbal WM task (i.e., *n*-Back task, Kane and Engle, 2002), which includes a manipulation component that requires participants to continuously update (every two seconds) their stored WM targets. Thus, to measure WM maintenance, we utilized a verbal *n*-Back task, which is known to be associated with DLPFC function and frontal theta activity (Gevins et al., 1997; Kane and Engle, 2002; Sandrini, Rossini, and Miniussi, 2008). The *n*-Back task is a cognitively demanding task forcing participants to actively maintain stimulus information during an inter-trial interval, as well as forcing them to inhibit the tendency to respond to a current stimulus (Krieger et al., 2005). This type of WM maintenance includes an executive attentional component that functions as the "gate keeper" of temporarily accessible information. During *n*-Back WM maintenance, the DLPFC's executive-attention component is constantly recruited to allow access to relevant to-be-remembered target-items in the presence of ongoing memory interference generated by intervening non-target items and previously retrieved items (Kane and Engle 2002). More so, in order to increase the task's sensitivity to the regulatory component of WM function, we implemented a Go-no-go response modality, which involves a response-inhibition subcomponent of executive attention considered to be associated with increased DLPFC activity (Kane and Engle, 2002; Meiron and Lavidor in press). The objective of the present study was to investigate transcranial alternating (i.e., oscillatory) current stimulation (tACS) effects (Antal et al., 2008; Schutter and Hortensius, 2011) on working memory (WM) maintenance and memory consolidation in healthy human participants.

Therefore, in response to Klimesch's (1999) hypothesis that theta synchronization is related to the encoding of new information, as well as his suggestion that long-term alterations in neural-network activity may be induced with stimulation patterns that mimic theta rhythm, we chose to use slow-wave-theta-band electric stimulation to presumably enhance endogenously-generated theta rhythm synchronization in the prefrontal cortex in order to investigate its effects on verbal memory maintenance and later episodic retrieval in healthy humans.

In a recent tACS study, healthy individuals displayed enhanced endogenous alpha band oscillations following 10 minutes stimulation over bilateral parieto-occipital locations in the alpha range (Zaehle, Rach, and Herrmann, 2010). These findings indicated that entrainment of band-specific oscillations may be used to modulate internal brain states in humans. Another study that investigated the effects of slow-oscillation brain stimulation on endogenous electroencephalography (EEG) rhythms and memory encoding revealed a transient time-dependent improvement in the encoding of verbal memory (immediate free recall of words) and a prominent increase in theta activity versus sham condition (Kirov et al., 2009). The enhanced theta-band power observed in Kirov et al.'s study (2009) supports the idea that incremental shifts in theta synchronization may be related to improved verbal memory maintenance.

In order to clarify the underlying neural-network mechanisms affected by non-invasive oscillatory current-stimulation, it has been suggested that neural circuits can be considered as neural oscillators in which the strength of their neural response is related to a spike-timing-dependent-plasticity induced by repetitive frequency-specific input resonating between neurons within a particular neural-network (Zaehle et al., 2010). Therefore, since multiple pathways lead from one neuron to others and back to the same neuron within a circuit, during phases of depolarization, theta-rhythm tACS might increase the likeliness of a neuron to fire in response to other neurons that fire or "resonate" in this particular frequency. tACS forces the membrane potential to oscillate away from its resting potential invoking network synchronization within a discrete frequency-specific circuit. This mechanism is known as a "stochastic resonance" neural-circuit response (Zaehle et al., 2010). In relevance, it is hypothesized that certain input frequencies selectively strengthen synapses that are incorporated in neural circuits that have a similar resonance frequency. Hence, prefrontal stochastic-resonance properties may underlie the predicted online WM enhancement during prefrontal theta frequency stimulation. In relevance to our experimental manipulation, the integrity of parietal-frontal network-connectivity has been implicated in having a central role in verbal WM maintenance (Karlsgodt et al., 2008). Additionally, Tulving et al.'s (1994) HERA model suggests increased left prefrontal cortical involvement in encoding verbal information into episodic memory. Thus, we included a left prefrontal-parietal stimulation condition.

In doing so, we were able to examine the benefits of left parietal-prefrontal tACS for online WM function, and for post-stimulation episodic retrieval versus a sham tACS condition. Thus, since frontal theta activity is related to higher brain functions (Koenig et al., 2001) such as the monitoring component of WM function (involved during increased demands on verbal memory maintenance), we predicted that online bilateral theta-stimulation to the prefrontal cortex of participants engaged in a verbal n-Back task will enhance online WM maintenance functions compared to both sham and active left prefrontal-parietal stimulation.

Furthermore, in support of the behavioral benefits of bilateral prefrontal theta synchronization for memory maintenance and implicit learning, we assessed whether bilateral prefrontal theta-stimulation will produce significantly higher post-stimulation WM accuracy than sham and the left parietal-prefrontal theta stimulation. Finally, in attempt to reveal theta stimulation long-term after-effects, we predicted higher post-stimulation episodic retrieval accuracy in both active stimulation conditions in comparison to a sham stimulation condition.

MATERIALS AND METHODS

Participants

A total of 35 female participants completed the study (ages 19-27, M = 21.62, SD = 1.84). We avoided the inclusion of males due to previous imaging and brain-stimulation findings that indicated gender differences in prefrontal lateralization during verbal WM performance (Goldstein et al., 2005; Meiron and Lavidor, in press; Speck et al., 2000). Twelve participants were in the active bilateral condition, 11 participants were in the active left condition, and 12 participants were in the sham condition. All participants fulfilled the following inclusion criteria: right handed, normal/corrected to normal vision and free of medication. The exclusion criteria were metallic implants, learning disabilities, neurological/psychiatric history, skin disease, first-degree relative suffering from a psychotic disorder/epilepsy, and pregnancy. All participants gave their informed consent. Prior to the tACS manipulation all participants completed a short-version standard handedness questionnaire (Oldfield, 1971) validating them as consistent right handers. The study was approved by the Ethics Committee of Bar-Ilan University.

Stimuli

Stimulus-displays presented 42 different pairs of word-nouns. Words within a word-pair were taken from different semantic categories (e.g., plants vs. animals). The words were one to two syllables long, and were translated to Hebrew and back to English by bi-lingual translators to verify that the original word meaning was kept in Hebrew. Lack of semantic associability between the words in each pair was confirmed using the Edinburgh Associative Thesaurus (Kiss et al., 1973) and empirical association data drawn from the *MRC Psycholinguistic Database* (Wilson, 1988). The word pairs were presented with font David, size 70. The words appeared simultaneously, at the top (center alignment, 25% -Y axis) and bottom (center alignment, 75% - Y axis) locations, both centered horizontally. Words were bound to a particular display-location to hypothetically induce greater event related theta synchronization during verbal WM maintenance (Wu et al., 2007). Stimuli were visually displayed on a computer screen using E-Prime v2.0 presentation software (Psychology Software Tools 2007). Screen size was 340 x 270 mm LCD.

Design

In order to construct a paradigm to investigate our hypothetical framework, target and intervening items that were presented during stimulation (*online* WM task) were again presented during the second post-stimulation WM task. The post stimulation WM task required cued recall in contrast to item-recognition in the first online WM task. Such a paradigm could reflect implicit learning effects (e.g., better performance on the second WM task) because the second WM task displays previously encountered target-items appearing within the same sequential verbal-context (e.g., the same WM trials). Furthermore, in order to investigate possible long-term plastic effects of active stimulation hypothesized to originate in output hippocampal-circuitry (Buzsaki, 1996; Chrobak and Buzsaki 1996; Cook and Bliss, 2006), we examined explicit episodic recall (i.e., conscious recollection, Rugg et al., 1998) 20 minutes after stimulation (third and final task) by asking subjects to freely recall all words that appeared as probes (e.g., single-word response displays, see next Materials and Methods subsection) during the first WM task. Additionally, we assessed implicit episodic recall (i.e., non-conscious recollection) by including the recollection of intervening non-target items in participants' WM accuracy scores.

Online Working Memory Task

Our verbal n-Back task was designed to recruit executive attention involvement (Kane and Engle, 2002) in two ways; first, we presented two possible targets within each n-Back study-display rather than one (see Figure 1), secondly, we used everyday words instead of letters. These particular interfering task-components have been shown to increase DLPFC activity (Kane and Engle, 2002) associated with executive attention's constant attempt to inhibit semantic retrieval-interference (Hicks and Starns, 2004) during item-recognition. Participants were instructed to study pairs of words within each two-second display for later 2-Back word recognition (e.g., correct recognition of an individual word that appeared within a pair of words two displays ago).

An individual trial ends when a single word-probe (e.g. response-display) appears at fixation, unexpectedly, following the presentation of two to five pairs of words (e.g., study displays). All stimuli-displays were spaced by inter stimulus intervals (ISI's) that directed participants to focus their attention on fixation cues in the center of the screen. In the response displays, participants were expected to respond/not respond by pressing/not pressing a key on the computer keyboard.

Therefore, following the presentation of a single word, participants were expected to respond by pressing a key only if they recognize the word as appearing two study-displays ago. If participants reject the word as a target, they were instructed not to respond. One trial included two to five study-displays (denoting four incremental levels of WM load, equally distributed across the entire task), ending with a two-second response-display presenting a single word in the center of the screen (see Figure 1). There were 12 different trials within one run of trials. The entire WM task consisted of four runs and five 30s inter-run resting periods (48 trials, approx 17 min long).

The four different runs displayed the same trials. The runs differed only in the presentation order of the trials and in their response-displays (probed-words). Within each run there were 12 different probed-words requiring the participants' response. 50% of the response displays were words taken from a two-back display (expected key-press response) and the other 50% were taken from non-target positions within a trial (e.g., one-Back or three-Back positions, no response expected).

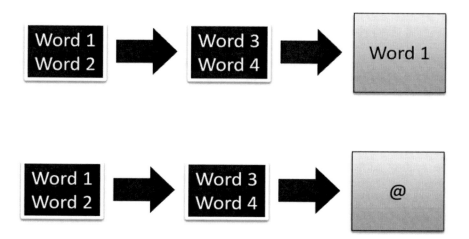

Figure 1. Top illustration: Online working memory task.

Illustration of a two study-display *n*-Back trial. Each trial begins with fixation (2000msec) and followed by two to five study-displays (2000msec each), interspaced by fixations, each displaying one word at the top and a different word at the bottom of the screen simultaneously. The last display within a trial presents one word in the middle of the screen for 2000msec (response-display) requiring the participant to press a key if it was part of a word-pair (word 1) that appeared two study-displays prior to response display. As illustrated, participants would be expected to press the key if word 1 or word 2 appeared in the response-display. Bottom illustration: Post-stimulation working memory task. Illustration of a two study-display *n*-Back trial. Each trial begins with fixation (2000msec) and followed by two to five study-displays (2000msec each), interspaced by fixations, each displaying one word at the top and a different word at the bottom of the screen simultaneously. The last display within a trial (response-display) presents one out of two possible symbols (@ or %) for 2000msec requiring the participant to recall either the top or the bottom word, respectively, that appeared two study-displays prior to response-display. As illustrated in the figure, the displayed trial required participants to recall Word 1(say out-loud the top-location word from two displays ago).

Online tACS Conditions

Following scalp measurements, participants were randomly assigned to three different "online" (stimulation during WM activity) tACS conditions. These conditions were 1) Active left prefrontal-parietal stimulation 2) Active bilateral DLPFC stimulation, and 3) Sham DLPFC stimulation. tACS was delivered by a battery driven Eldith DC-stimulator (neuroConn GmbH, Germany) using a pair of conductive-rubber electrodes placed in saline-soaked synthetic sponge. Both electrodes were 4 x 4 cm. All electrode positions were in accordance with the international 10-20 EEG system. The tACS bipolar montage for the active bilateral prefrontal condition was placed over the left and right DLPFC (corresponding to F3/AF3 midpoint and F4/AF4 midpoint EEG electrode locations, respectively). Both left and right prefrontal stimulation sites were located according to Fitzgerald et al.'s (2009) findings for optimal localization of the DLPFC (i.e., midpoint corresponding to halfway between F3 and AF3 electrode locations).

The left unilateral bipolar electrode montage was placed over the P1/P3 midpoint (left supramarginal gyrus, Herwig, Satrapi, and Schönfeldt-Lecuona, 2003) and F3/AF3 midpoint electrode locations. The sham stimulation montage was the same as the active bilateral DLPFC montage. Alternating current stimulation was sinusoidal and current intensity was set to 1000μA producing a peak current intensity of ± 500μA. Alternating current frequency was set at 4.5Hz. These stimulation parameters were predetermined following the assertion that they did not produce phosphenes or persisting skin sensation in a preliminary testing session with five healthy participants. 5400 cycles (20 minutes stimulation) were delivered in the active conditions. Current was ramped up and down for 10 seconds and impedance was kept below 10 kΩ. In the sham stimulation, all parameters were the same as the active conditions, however only 90 cycles of stimulation were delivered (20 seconds). Following all experimental procedures, a debriefing was conducted to find out if participants felt the stimulation.

Post-Stimulation Verbal WM Task

In order to investigate short-term after-effects of the online tACS manipulation, immediately after completing the online task, participants were required to complete a similar verbal n-Back task that examined implicit learning effects by measuring cued recall instead of item recognition (see Figure 1 for illustration). The post-stimulation WM task required explicit verbal recall (total of 48 target-words), which is associated with successful context-dependent consolidation into long-term verbal memory, a behavioral byproduct of executive attention "active maintenance" WM function (Kane and Engle 2002). The target and study items were the same as in the online WM task; however, in the response displays, one out of two possible abstract cues was presented, which directed the participant to explicitly recall a location-specific item that appeared two displays ago. Vocal responses were recorded during the 2-second response displays by the computer.

Post-Stimulation Free Recall Task (Episodic Memory)

In order to examine long-term behavioral effects, following the post stimulation cued recall WM task (20 minutes post stimulation), participants were instructed to freely recall all the probe-words (e.g., response displays) that appeared in the first WM task. Participants had 90 seconds to say out loud all the probed single-words they remembered (i.e., explicit recall) from the first WM task (from a total of 48 probe-words). They were encouraged to recall as many words as possible (from a total of 84 words), including words with a low confidence level (e.g., guessing).

This measure reflects implicit recall because participants are also unconsciously recalling words form the context of the probed target-words, while making a conscious effort to remember the probed target-words from the first WM task. The computer recorded their vocal responses.

Post Stimulation Subjective Measures

As a standard indication of the participants' level of discomfort, following the free recall post-stimulation task, participants were asked to rate the pain-intensity they endured during the stimulation using a visual analogue scale (VAS) (Jensen, Karoly, and Braver, 1986). Participants were also questioned about their stimulation experience (e.g., "how do you feel after this experience?", "what did you feel during stimulation?").

Procedure

The entire online tACS session was 20 minutes long. We employed a single-blind procedure informing participants that they would receive active stimulation. During the first three minutes of stimulation participants performed a practice run, including a verification procedure that they have understood the task. In the remaining 17 minutes of tACS, participants performed a verbal *n*-Back task (first task) while receiving one out of three possible tACS manipulations - left prefrontal-parietal stimulation, bilateral prefrontal stimulation, or sham stimulation. A between-subject design was preferable since performing the same memory challenges twice (as would be expected in a within-subject design) is likely to produce confounding carry-over learning effects, which are not associated with online WM executive attention involvement (Kane and Engle, 2002). Participants' head positions were maintained at a distance of 57cm from the center of the screen using a chin-rest apparatus. Following the online WM task, participants had the electrode montage removed from their head (3 minutes) and immediately began the post-stimulation WM task (17 min). The post-stimulation WM task (second task) was followed by a 90 seconds free recall task (third and final task). Finally, following the three consecutive memory challenges, participants were interviewed for three to five minutes and their subjective reports were recorded by the experimenter.

Working Memory and Episodic Memory Measures

Response-accuracy (proportion of correct responses) was used as the main dependent variable for later statistical analyses. Reaction times for correct responses on the first WM task were also included in the analyses. Episodic memory accuracy was measured in two ways: 1) the number of words recalled that matched words that appeared during the online WM task (implicit episodic memory accuracy), and, 2) the number of words recalled that appeared only in the response displays during the online WM task (explicit episodic memory accuracy).

We discriminated between these two measures to investigate whether tACS manipulations influenced implicit recall versus explicit recall, since explicit recall of context-dependent events is more dependent on long-term plastic changes in the hippocampus (Cook and Bliss, 2006).

Data Analysis

The overall percentage of correct responses from the online and post-stimulation cognitive tasks was entered into a 3-level (active left, active bilateral and sham condition) one-way analysis of variance (ANOVA).

Further, we evaluated the effects of tACS condition on post-stimulation WM accuracy and post-stimulation episodic-retrieval accuracy. Bonferroni tests were used for post hoc comparisons with $p < 0.05$.

RESULTS

The stimulation was well tolerated (as indicated by extremely low VAS scores, see Table 1), there were no side effects, and participants' reports indicated they could not discriminate between the active versus the sham stimulation condition (forced guessing was at chance level).

Table 1. Table data represents 35 female participants. Hand variable represents right handedness scores (from 0 to 100), Education variable represents the number of years in school, VAS represents pain intensity ratings post-stimulation on a scale from 0 to100 millimeters. SD represents standard deviation

	Active left		Sham		Active bilateral	
	Mean	SD	Mean	SD	Mean	SD
Hand	72.27	10.00	74.45	8.04	72.71	9.00
Education	12.27	.90	12.42	1.00	12.67	1.07
Age	21.91	1.30	21.50	2.43	21.50	1.73
VAS	5.55	8.43	1.58	3.80	.83	1.40

There were no significant differences in the pain-intensity ratings between the three tACS conditions ($F(2, 32) = 2.6$, $p = .089$). In order to control for group homogeneity, all participants were female, and participants in the three different tACS conditions were matched for age, education, and handedness (see Table 1).

Online WM Accuracy

A one-way ANOVA with three levels of tACS manipulation (left active vs. bilateral active vs. sham stimulation) was employed to examine tACS condition effects on online WM accuracy. This analysis showed a main effect of tACS condition ($F(2, 34) = 4.98$, $p = .013$). Bonferroni post hoc tests (with corrected $p < .05$) showed that the active bilateral DLPFC stimulation ($M = .93$, $SD = .03$) induced significantly higher WM accuracy than the active left condition ($M = .85$, $SD = .07$), as well as significantly higher (see Figure 2) accuracy than the sham condition ($M = .87$, $SD = .06$). There was no significant tACS effect on reaction times for correct responses.

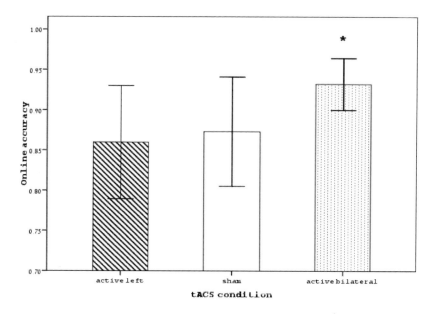

Figure 2. Enhanced online working memory accuracy in the active bilateral DLPFC condition. Error bars display +/- 1 SD, * significantly different from sham ($p = .03$) and active left ($p = .01$) conditions. The Y axis represents the proportion of correct responses during the online WM task.

Post-Stimulation WM Accuracy

To examine the effect of tACS stimulation on post-stimulation WM accuracy a one-way ANOVA (Sham vs. Active left vs. Active bilateral) was performed. This analysis yielded no main effect of tACS condition on post-stimulation WM accuracy ($F(2, 33) = 2.49$, $p = .09$). Using Pearson correlation, we examined the relationship of online WM accuracy and post-stimulation WM accuracy within each tACS condition and found a significant relationship between online WM accuracy and post stimulation WM accuracy only in the sham condition ($r = .61$, $p = .03$).

Post-Stimulation Episodic Retrieval Accuracy (Free Recall Task)

We found a main effect of tACS condition ($F(2, 31) = 5.33$, $p = .01$) on implicit episodic memory accuracy (IM accuracy) scores. Bonferroni post hoc corrections ($p < .05$) indicated that the active bilateral condition produced significantly higher IM accuracy ($M = .18$, $SD = .07$) than sham stimulation ($M = .10$, $SD = .05$). Although post hoc significance level did not reach (corrected for multiple comparisons) significance in the left active condition ($p = .048$), the left active condition also produced higher IM accuracy ($M = .17$, $SD = .06$) than the sham condition. Next, we examined explicit episodic memory accuracy (EXM accuracy) and found a significant main effect of tACS condition ($F(2, 31) = 4.27$, $p = .02$). Bonferroni post hoc corrections ($p < .05$) indicated that only the active bilateral condition ($M = .25$, $SD = .1$)

produced significantly higher EXM accuracy versus sham condition (M = .14, SD = .09, Figure 3).

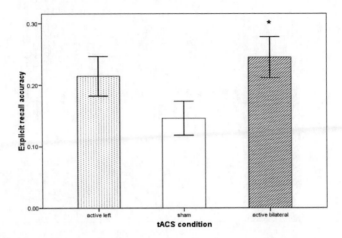

Figure 3. Greater explicit recall accuracy (20 min post stimulation) in the active bilateral DLPFC condition versus sham condition. Error bars display +/- 1 SE, * significantly different from sham ($p = .02$). Y-axis represents *explicit* episodic memory accuracy (EXM accuracy).

As a result of relatively small sample sizes we chose to present the distribution of the accuracy scores in the three different verbal memory challenges. Figure 4 presents the distribution of accuracy scores for the online WM task, post stimulation WM task (3 minutes after stimulation), and episodic recall scores attained 20 minutes post stimulation (explicit recall as well as implicit recall accuracy).

DISCUSSION

Our results highlight the importance of bilateral prefrontal involvement in WM retrieval and episodic memory encoding. These long-term behavioral effects may support our hypothesis that bilateral theta-rhythm weak-current stimulation to the DLPFC during executive WM maintenance may induce long-term network-plasticity effects in hippocampal-prefrontal output circuitry involved in context-dependent learning (Buzsaki 1996; Chrobak and Buzsaki 1996; Cook and Bliss, 2006; Wang, 2010). However, further investigation is needed to examine the effect of other frequency bands (i.e., alpha-band frequency), as well as the effect of unilateral right parietal-DLPFC stimulation, in order to confirm the observed theta-stimulation enhancement. Pursuing the study of other frequency-band stimulations may present a methodological challenge as our earlier preliminary tests indicated either phosphene-induction (peripheral rhythmic flashes of light) and/or somatosensory sensations (rhythmic tactile sensation) in the subjective reports of participants who received alpha (12 Hz), beta (15 Hz), and gamma-band (40 Hz) frequency stimulation with the same prefrontal montage and current peak-intensity. In conjunction, we suggest that it would also be of great interest to examine only male participants with the same experimental paradigm as recent

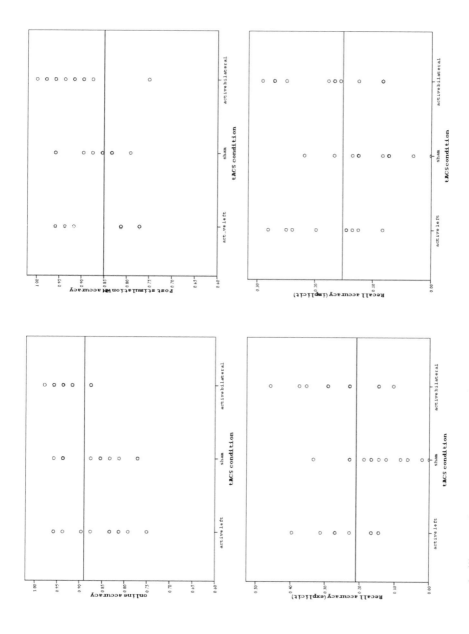

Figure 4. The two top graphs illustrate the accuracy-score-distribution as a function of tACS condition for the online WM task (left side) and post stimulation WM task (right side). The two bottom graphs illustrate accuracy-score-distribution as a function of tACS condition for the free recall task 20 minutes post stimulation. Explicit episodic recall accuracy is presented on the left side, and implicit episodic recall accuracy is presented on the right side. The horizontal line in each illustration represents the mean accuracy score across all participants.

tDCS findings have shown that prefrontal stimulation effects on WM function are modulated by gender (Meiron and Lavidor, in press). Importantly, to our knowledge, this is the first study that shows tACS after-effects (20 minutes post-stimulation) on explicit memory recall in healthy humans.

As predicted, online WM accuracy was significantly higher in the active bilateral condition than the left active condition and the sham condition. This finding coincides with our theoretical premise stipulating that bilateral DLPFC theta synchronization facilitates verbal WM maintenance versus unilateral parietal-DLPFC theta synchronization. Interestingly, this finding may compliment Wu et al.'s (2007) EEG results, which revealed that working memory binding involves large-scale neural synchronization at the theta frequency range. In relevance, since we have assessed the monitoring component of WM maintenance our findings could also support the theoretical perspective that theta oscillations play an important role in top-down processing during WM retrieval (Wang, 2010). In support, in a previous study we employed the same paradigm but with anodal non-oscillatory transcranial direct current stimulation (tDCS) and found no effect on overall accuracy scores (Meiron and Lavidor, in press). Our previous tDCS findings implied that unilateral anodal tDCS to the either left or right DLPFC during verbal WM activity might improve accuracy only on high-load WM items. Therefore, overall WM retrieval enhancement, as observed in the present study, could be dependent on long-distance theta-band neural synchronization. These theta synchronization benefits may reflect increased functional connectivity (Koenig et al., 2001), and mutual bilateral interdependence of the DLPFC's during verbal WM maintenance. Additionally, correlation analyses indicated that online WM accuracy was positively correlated with post-stimulation WM accuracy only in the sham condition. This finding indicated that the active tACS conditions might have altered DLPFC function, producing WM scores that were unrelated to WM scores in the post-stimulation WM task. Finally, although previous EEG data (Koenig et al., 2001; Wang, 2010; Wu et al., 2007) support the critical role of bilateral prefrontal theta synchronization during WM performance, still, further investigation that compares other stimulation frequencies and their effect on EEG band-specific power is needed to fully support our findings. Although we have observed significant online tACS effects on WM function, we did not find significant tACS effects on post-stimulation WM performance. This null finding was also observed in a previous study using anodal tDCS over the left or right DLPFC (Meiron and Lavidor, in press). Thus, we suggest that including a similar WM task following the initial online WM task may have impeded on the DLPFC's sensitivity to the preceding excitatory stimulation. Hence, during the post-stimulation WM task, which required participants to recall WM items that have already appeared within the online WM task, the DLPFC's propensity for modulation may have been arrested. Therefore, implicit learning of WM targets may have been maximized in all tACS conditions as a result of repeated exposure to the same WM trials. This notion may support the null tACS effects in the post stimulation WM task.

Critical to our investigation, we have showed that both implicit and explicit episodic memory retrieval were significantly enhanced in the active tACS conditions. Most importantly, these effects were detected in the free recall task, 20 minutes post tACS session. Interestingly, implicit episodic retrieval was positively affected by the left parietal-DLPFC stimulation, as well as significantly improved in the bilateral DLPFC stimulation. This finding partially supports the HERA model (Tulving et al., 1994), which predicts increased left prefrontal involvement during verbal-episodic encoding. In the present study, the left stimulation advantage over sham is associated with increased implicit retrieval of episodic memory, but this effect was found to be significant only in the active bilateral DLPFC stimulation condition. Therefore, implicitly recalling verbal items may rely on increased left parietal- DLPFC activity, and alternatively, may be significantly facilitated by increased bilateral DLPFC activity. These findings imply a more diffused cortical network involvement (left parietal-DLPF as well as bilateral DLPFC networks) in implicit memory processes. In support, Tulving et al. (1994) have suggested that implicit verbal memory retrieval is associated with activity in several posterior sites, increased activation in the right DLPFC, as well as increased involvement of left frontal areas along the medial prefrontal cortex. Hence, explicit memory processes may be dependent on a less cortically-diffused, more discrete cortico-hippocampal circuitry involving a dialogue (i.e., increased theta band synchronization) between the hippocampus and the prefrontal cortex (Buzsaki, 1996). In relevance, we believe that our bilateral prefrontal theta stimulation may have engaged large-scale network theta-oscillations, inducing "intrinsic resonant properties" (Buzsaki, 1996) within hippocampal networks that are sensitive to the theta frequency when attempting to achieve successful memory consolidation. This neurophysiological interpretation is supported by recent findings that indicated that enhanced memory performance is predicted by theta-frequency spike field coherence in the output circuitry of the hippocampus (Rutishauser et al., 2010). Furthermore, our findings could support the theoretical standpoint, which states that human cortical theta rhythm serves to coordinate or "synchronize" activity between the hippocampus and cortical areas (Caplan et al., 2003). As indicated in the bilateral DLPFC stimulation condition, hypothetically, increased theta synchronization between the hippocampus and the DLPFC during executive WM maintenance may have induced long-term plastic changes needed to consolidate WM events into long-term episodic memory.

Importantly, the observed bilateral-DLPFC tACS enhancement seems to support Klimesch's (1999) suggestion that long-term plastic effects may be induced with stimulation patterns that mimic theta rhythm. However, although our bilateral theta stimulation promoted long-term behavioral effects reflected in enhanced explicit episodic recall, future electrophysiological investigations using prefrontal-connectivity EEG data-analysis (i.e., theta-band *Global Field Synchronization*, see Koenig et al., 2001) during post-stimulation free recall could support or disconfirm the induction of long-term prefrontal-network theta-activity changes following DLPFC theta-band stimulation. Presumably, our bilateral DLPFC tACS manipulation may have enhanced endogenously-generated theta rhythm synchronization in the prefrontal cortex.

Memory-consolidation, as indicated by significantly higher explicit recall scores in the active bilateral condition versus sham condition, is hypothesized to be associated with long-term plastic changes in limbic structures (i.e., hippocampus) and interconnected prefrontal areas (Buzsaki 1996; Caplan et al., 2003; Chrobak and Buzsaki, 1996; Cooke and Bliss, 2006; Griesmayr et al., 2010; Paz et al., 2011).

In sum, our behavioral results show promise for cognitive remediation to be used as an adjunctive technique with tACS. The online WM enhancement observed in the active bilateral condition demonstrates that DLPFC theta stimulation may be used to increase transient functional connectivity during WM maintenance.

The post-stimulation episodic memory enhancement observed in the active bilateral condition implies that DLPFC theta stimulation may contribute to long-term changes within hippocampal-prefrontal episodic networks. In relevance, our findings indicate that our bilateral DLPFC stimulation protocol may be used to increase functional connectivity related to theta-band synchronization in clinical populations that suffer from loosened theta-band functional connectivity. For instance, first-episode schizophrenia patients display reduced theta band synchronization (Koenig at al., 2001), which is hypothesized to generate dissociative memory contents.

Furthermore, Wang (2010) and Koenig et al. (2001) suggested that schizophrenia patients suffer from impairments in long-range connections throughout the entire brain, which are manifested by an array of cognitive deficits such as distractibility of attention and WM dysfunction. Accordingly, our main results provide preliminary evidence for the feasibility of bilateral DLPFC tACS to be used as a therapeutic application to enhance WM function in schizophrenia patients.

Finally, our theta-frequency tACS after-effects findings (e.g., enhanced explicit recall) are in consonance with contemporary neurophysiological learning models implicating theta-phase modulation in memory formation and retrieval (Buzsaki 1996; Cooke and Bliss 2006; Wang, 2010). Future studies examining the effect of other band-specific stimulation protocols in conjunction with electrophysiological data collection are needed to support the specific contribution of bilateral prefrontal theta stimulation to WM function and episodic retrieval.

ACKNOWLEDGMENTS

This study was supported by an ERC startup grant awarded to ML (Inspire 200512). The authors would like to thank cognitive neuroscience lab (Bar Ilan University) research team for helpful discussion.

REFERENCES

Antal A, Boros K, Poreisz C, Chaieb L, Terney D, Paulus, W (2008). Comparatively weak after-effects of transcranial alternating current stimulation (tACS) on cortical excitability in humans. *Brain Stimulation* 2: 97-105.

Baddeley AD (1986). *Working memory.* Oxford University Press, Oxford.

Buzsaki G. (1996). The hippocampo-neocortical dialogue. *Cereb Cortex* 6:81-92.

Caplan JB, Madsen JR, Schulze-Bonhage A, et al (2003). Human θ oscillations related to sensorimotor integration and spatial learning. *J. Neurosci.* 23(11):4726-4736.

Chrobak JJ, Buzsaki G (1996). High-frequency oscillations in the output networks of the hippocampal-entorhinal axis of the freely behaving rat. *J. Neurosci.* 16(9):3056-3066.

Cooke SF, Bliss TVP (2006). Plasticity in the human central nervous system. *Brain* 129:1659-1673.

Fitzgerald PB, Maller JJ, Hoy KE, et al (2009). Exploring the optimal site for the localization of dorsolateral prefrontal cortex in brain stimulation experiments. *Brain stimulation* 2:234-237.

Gevins A, Smith ES, McEvoy L, et al (1997). High-resolution EEG mapping of cortical activation related to working memory:Effects of task difficulty, type of processing, and practice. *Cereb Cortex* 7:374-385.

Goldstein JM, Jerram M, Poldrack R, et al (2005). Sex differences in prefrontal cortical brain activity during fMRI of auditory verbal working memory. *Neuropsychology* 19:509-519.

Griesmayr B, Gruber WR, Klimesch W, et al (2010). Human frontal midline theta and its synchronization to gamma during a verbal delayed match to sample task. *Neurobiol. Learn Mem.* 93:208-215.

Hicks JL, Starns JJ (2004). Retrieval-induced forgetting occurs in tests of item recognition. *Psychon. Bull. Rev.* 11(1):125-130.

Herwig U, Satrapi P, Schönfeldt-Lecuona C (2003). Using the International 10-20 EEG System for Positioning of Transcranial Magnetic Stimulation. *Brain Topography* 16:95-99.

Jensen MP, Karoly P, Braver S (1986). The measurement of clinical pain intensity: a comparison of six methods. *Pain* 27(1):117-126.

Kane MJ, Engle RW (2002). The role of prefrontal cortex in working-memory capacity, executive attention, and general fluid intelligence: an individual-differences perspective. *Psychon. Bull. Rev.* 9:637-671.

Karlsgodt KH, van Erp TGM, Poldrack RA, et al (2008). Diffusion tensor imaging of the superior longitudinal fasciculus and working memory in recent-onset schizophrenia. *Biol. Psychiatry.* 63:512-518.

Kiss GR, Armstrong C, Milroy R, et al (1973). An associative thesaurus of English and its computer analysis. In: Aitken AJ, Bailey RW, and Hamilton-Smith N, (eds) The Computer and Literary Studies. Edinburgh: University Press: pp. 153-165.

Koenig T, Lehmann D, Saito N, et al (2001). Decreased functional connectivity of EEG theta-frequency activity in first-episode, neuroleptic-naïve patients with schizophrenia: preliminary results. *Schizophr. Res.* 50:55-60.

Krieger S, Lis S, Cetin T, et al (2005). Executive function and cognitive subprocesses in first-episode, drug-naive schizophrenia: An Analysis of N-Back Performance. *Am. J. Psychiatry* 162:1206–1208.

Kirov R. Weiss C, Siebner HR, et al (2009). Slow oscillation electrical brain stimulation during waking promotes EEG theta activity and memory encoding. *Proc. Natl. Acad. Sci. USA* 106:15460-15465.

Klimesch W. (1999). Alpha and theta oscillations reflect cognitive and memory performance: a review and analysis. *Brain Res. Rev.* 29:169-195.

Meiron O, Lavidor M (in press). Unilateral prefrontal direct current stimulation effects are modulated by working memory load and gender. *Brain Stimulation.*

Oldfield RC (1971). The assessment and analysis of handedness: the Edinburgh inventory. *Neuropsychologia*, 9(1), 97-113.

Paz R, Bauer PE, Pare D (2011). Theta synchronizes the activity of medial prefrontal neurons during learning. *Learn memory* 15:524-531.

Rugg MD, Ruth EM, Walla P. et al (1998). Dissociation of the neural correlates of implicit and explicit memory. *Nature* 39:595-597.

Rutishauser U, Ross IB, Mamelak AN, et al (2010). Human memory strength is predicted by theta-frequency phase-locking of single neurons. *Nature* 464:903-907.

Sandrini M, Rossini PM, Miniussi C (2008). Lateralized contribution of prefrontal cortex in controlling task-irrelevant information during verbal and spatial working memory tasks: rTMS evidence. *Neuropsychologia* 46:2056-2063.

Schack B, Vath N, Petsche H, et al (2002). Phase-coupling of theta-gamma EEG rhythms during short-term memory processing. *Int. J. Psychophysiol.* 44:143-163.

Schutter DJ, Hortensius R (2011). Brain oscillations and frequency-dependent modulation of cortical excitability. *Brain Stimulation* 4:97–103.

Speck O, Ernst T, Braun J, et al (2000). Gender differences in the functional organization of the brain for working memory. *Neuroreport* 11:2581-2585.

Tulving E, Kapur S, Craik FI, et al (1994). Hemispheric encoding/retrieval asymmetry in episodic memory: positron emission tomography findings. *Proc. Natl. Acad. Sci. USA* 91:2016-2020.

Wang X-L (2010). Neurophysiological and computational principles of cortical rhythms in cognition. *Physiol. Rev.* 90:1195-1268.

Wilson MD (1988). The MRC Psycholinguistic Database: Machine Readable Dictionary, Version 2. *Behav. Res. Meth. Instrum. Comput.* 20(1):6-11.

Wu X. Chen X. Li Z. et al (2007). Binding of verbal and spatial information in human working memory involves large-scale neural synchronization at theta frequency. *NeuroImage* 35:1654-1662.

Zaehle T, Rach S, Herrmann CS (2010). Transcranial alternating current stimulation enhances individual alpha activity in human EEG. *PLoS ONE* 5(11):e13766.

Chapter 12

COGNITIVE REMEDIATION OF WORKING MEMORY IN ADHD

Robert M. Roth and Shanna Treworgy
Department of Psychiatry, Geisel School of Medicine at Dartmouth, Lebanon, NH, US

ABSTRACT

Deficient working memory has been reported to be a core cognitive abnormality in attention-deficit / hyperactivity disorder (ADHD), and it has been associated with negative outcomes in a variety of areas such as academic and social functioning. Despite the salience of working memory deficits to ADHD, few interventions have been developed that seek to directly rehabilitate this cognition function. Cognitive remediation has been receiving increasing empirical attention as a means to directly ameliorate working memory in ADHD. In the present chapter, we review research that has employed computerized and other remediation approaches to improve working memory in this population. Findings pertaining to effects of training on near transfer, far transfer, and the core symptoms of ADHD are discussed. Methodological issues are considered including sample characteristics, variability in remediation programs and outcome measures, as well as the timeframe in which maintenance of treatment gains were assessed. Overall, cognitive remediation appears to be a promising approach for improving working memory in ADHD, though further research is required.

INTRODUCTION

Attention deficit hyperactivity disorder (ADHD) is a developmental disorder first manifesting in childhood, and characterized by pervasive symptoms of inattention, impulsivity and hyperactivity. It has been reported to affect 3-7% of school-aged children (American Psychiatric Association, 2000), and may persist into adulthood for as many as 50 to 66% of these children (Barkley, Fischer, Smallish, and Fletcher, 2002; Lara et al., 2009). In adult samples, epidemiological research has observed a prevalence rate of 4.4% in the U.S. (Kessler et al., 2006), while a rate of 3.4% was observed in the cross-national DSM–IV World Health Organization World Mental Health Survey Initiative (Fayyad et al., 2007).

These findings are especially meaningful when considered in the context of the multiple negative outcomes related to the disorder. In children and adolescents, ADHD has been associated with problems such as impaired academic performance (Hinshaw, 2002; Massetti et al., 2008) and social functioning (Hinshaw, 2002), and reduced quality of life (Danckaerts et al., 2010). Adults with ADHD are reported to have lower educational and occupational attainment (Klein et al., 2012; Kuriyan et al., 2013), more workplace-related injuries (de Graaf et al., 2008; Kessler, Lane, Stang, and Van Brunt, 2009), high rates of marital instability (Klein et al., 2012; Murphy and Barkley, 1996), reduced quality of life (Agarwal, Goldenberg, Perry, and Ishak, 2012), as well as more frequent risky driving behaviors and driving-related accidents (Barkley, Murphy, Dupaul, and Bush, 2002; Richards, Deffenbacher, and Rosen, 2002; Thompson, Molina, Pelham, and Gnagy, 2007).

Defining Working Memory

Working memory has been emphasized in many theoretical models and the focus of numerous empirical investigations in clinical disorders such as ADHD. Although there are a number of models of working memory (see Miyake and Shah, 1999), Baddeley's model has been the most commonly cited in relation to ADHD (Baddeley, 1992, 1996). This model describes working memory as a multicomponent system that supports executive control and the active maintenance of information in mind. Information is stored in limited capacity, temporally limited verbal and spatial storage buffers, also referred to as the phonological or articulatory loop (which holds and maintains acoustic or speech-based information) and the visuospatial sketchpad (which holds and maintains visual images). More recently, an additional component was added to the model, the episodic buffer, which temporarily stores information that is multidimensional in nature (i.e., combining auditory and visual information, as well as potentially information from other sensory modalities), and allows for the interaction of the various components of working memory as well as information from long-term memory and perception (Baddeley, 2010). These buffers are regulated by a "central executive" component of the system which manages the distribution of limited attentional resources, and actively maintains and coordinates information within the buffers. More generally, working memory may be thought of as a cognitive process that actively holds and manipulates information in mind for the purpose of goal-directed behavior.

Evidence for Working Memory Deficit in ADHD

A number of neuropsychological models have emphasized the role of working memory in ADHD. Barkley's influential model stresses the importance of both behavioral inhibition and working memory, but views deficits in inhibition as the core disturbance that contributes to working memory and other executive deficits (Barkley, 1997a, 1997b). In contrast, Rapport and colleagues argue that a working memory deficit is likely the core abnormality in the disorder, with inhibitory and other executive deficits being secondary to it (Rapport, Chung, Shore, and Isaacs, 2001). This latter view echoes prior assertions that working memory is "the hub of cognition" (Haberlandt, 1997; p. 212) and "perhaps the most significant achievement of human mental evolution" (Goldman-Rakic, 1992; p. 111).

Studies employing a wide variety of performance-based measures have provided considerable evidence for the presence of working memory deficit in children, adolescents, and adults with ADHD (e.g. Alloway, 2011; Berlin, Bohlin, Nyberg, and Janols, 2004; Brocki, Randall, Bohlin, and Kerns, 2008; Dowson et al., 2004; Finke et al., 2011; O'Brien, Dowell, Mostofsky, Denckla, and Mahone, 2010; Raiker, Rapport, Kofler, and Sarver, 2012; Toplak, Bucciarelli, Jain, and Tannock, 2008). There is also some support for the primacy of working memory deficit over behavioral disinhibition in the disorder (Alderson, Rapport, Hudec, Sarver, and Kofler, 2010; Raiker et al., 2012), though literature directly addressing this issue is sparse. Support for the presence of deficits on working memory tasks in ADHD has also been provided by meta-analyses (Kasper, Alderson, and Hudec, 2012; Martinussen, Hayden, Hogg-Johnson, and Tannock, 2005; Willcutt, Doyle, Nigg, Faraone, and Pennington, 2005). While some studies have suggested that comorbid diagnoses (e.g., learning disorder) may contribute to the working memory deficit in ADHD (Jonsdottir, Bouma, Sergeant, and Scherder, 2005; Katz, Brown, Roth, and Beers, 2011), meta-analytic studies have concluded that the deficit is independent of comorbid disorders as well as general intellectual ability (Martinussen et al., 2005; Willcutt, et al., 2005), and that variability among studies cannot be accounted for by the source of clinical samples (community or clinic) or which diagnostic classification system was employed (Willcutt et al., 2005). Additional support for working memory being a core deficit in ADHD comes from evidence that unaffected siblings of children with the disorder also demonstrate difficulties with working memory (Rommelse et al., 2008).

The working memory impairment in ADHD may be more prominent for spatial than verbal tasks (Rhodes, Park, Seth, and Coghill, 2012), though findings have been inconsistent, possibly due to discrepancies in relative task difficulty (Brocki et al., 2008). Age of study samples may also play a role, verbal deficits being more likely to be observed in younger than older children with the disorder (Sowerby, Seal, and Tripp, 2011). The specific components of working memory examined have also been found to be relevant, with deficits seen in both storage (Finke et al., 2011; Martinussen and Tannock, 2006; Rhodes et al., 2012) and the central executive (Martinussen and Tannock, 2006; Rhodes et al., 2012). Rapport and colleagues found that children with ADHD show deficits in three aspects of working memory identified using a latent variable approach (phonological buffer/rehearsal loop, visuospatial buffer/rehearsal loop, and the central executive attentional controller), with the most prominent deficit being observed in the central executive (Rapport et al., 2008). These findings were observed even after controlling for age, IQ, socioeconomic status, reading speed, and nonverbal visual encoding.

Importantly, working memory deficits in ADHD have been not only identified on laboratory-based measures, but also ratings scales of everyday behavior such as the Behavior Rating Inventory of Executive Function (BRIEF; Gioia, Isquith, Guy, and Kenworthy, 2000) and the Working Memory Rating Scale (Alloway, Gathercole, and Elliott, 2010). This has been observed in preschoolers (Mahone and Hoffman, 2007), older children (Jarratt, Riccio, and Siekierski, 2005; Mahone et al., 2002; McCandless and O' Laughlin, 2007), adolescents (Toplak et al., 2008), and adults (Rotenberg-Shpigelman, Rapaport, Stern, and Hartmen-Maeir, 2008; Roth, Isquith, and Gioia, 2005) with the disorder. For example, Gioia and colleagues (Gioia et al., 2000) reported that the BRIEF Working Memory scale predicted ADHD-Inattentive type (ADHD-I) and ADHD-Combined type (ADHD-C) with approximately 80% accuracy for both the Parent and Teacher form. These findings indicate that problems with working memory in individuals with ADHD may also be observed in their everyday environments such as in the classroom, at work, and at home.

Finally, several lines of evidence indicate that working memory deficits contribute to functioning difficulties in the everyday environment. For example, in ADHD poorer working memory has been associated with worse social functioning (Kofler et al., 2011) and academic achievement (Alloway et al., 2010; Gropper and Tannock, 2009; Rogers, Hwang, Toplak, Weiss, and Tannock, 2011). This is consistent with studies of other populations demonstrating the impact of working memory on functional outcomes such as social and emotional functioning (Clark, Prior, and Kinsella, 2002), academic achievement (Gathercole, Alloway, Willis, and Adams, 2006; Lan, Legare, Ponitz, Li, and Morrison, 2011), and adaptive behavior (Gilotty, Kenworthy, Sirian, Black, and Wagner, 2002).

APPROACHES TO IMPROVING WORKING MEMORY IN ADHD

As noted above, there is substantial evidence for the presence and importance of working memory deficits in ADHD. The majority of treatments, however, were developed to target the core symptoms of inattention, impulsivity and hyperactivity rather than working memory. Furthermore, despite the growing interest in cognitive enhancers for ADHD (Bidwell, McClernon, and Kollins, 2011), relatively few studies have directly examined the effects of treatment on working memory in individuals with the disorder.

Medication

A number of medications that are effective at reducing the core symptoms of ADHD (i.e., inattention, impulsivity and hyperactivity) have also been reported to show promise for ameliorating working memory in those with the disorder. Several studies have reported that treatment with methylphenidate is associated with improvements in working memory on performance-based tests (e.g. Mehta, Goodyer, and Sahakian, 2004; Rubio, Hernandez, Verche, Martin, and Gonzalez-Perez, 2011; Turner, Blackwell, Dowson, McLean, and Sahakian, 2005; Yang et al., 2011), as well as on parent and teacher ratings on the BRIEF (Yang et al., 2011); though negative findings have also been reported (e.g., Blum, Jawad, Clarke, and Power, 2011). This inconsistency may be related to several factors such as the nature of the measures employed (Bedard, Jain, Johnson, and Tannock, 2007), possible differential effects of methylphenidate on the components of working memory assessed (Bedard and Tannock, 2008), the duration of treatment (Rubio et al., 2011), and the presence of comorbid conditions such as anxiety (Bedard and Tannock, 2008).

Atomoxetine, a non-stimulant medication for ADHD, has also been found to result in better working memory on performance-based tests (Gau and Shang, 2010; Sumner et al., 2009; Yang et al., 2011) and improved working memory in everyday life as assessed using parent/teacher ratings scales (Maziade et al., 2009; Yang et al., 2011). Recently, memantine, an N-methyl-d-aspartate (NMDA) receptor antagonist, has been reported to improve working memory in an open label study of adults with ADHD (Surman et al., in press). Despite these promising findings, reviews of the literature have suggested that the effects of currently available medications for ADHD have limited, albeit generally positive, effects on working memory for those with the disorder. A meta-analysis of methylphenidate studies indicated a significant effect in only 4 out of 8 studies, with higher doses having a more pronounced effect (Pietrzak, Mollica, Maruff, and Snyder, 2006); although tasks included in the analysis involved those tapping working memory as well as other executive functions. In a review of controlled trials of stimulants in ADHD, Swanson and colleagues (Swanson, Baler, and Volkow, 2011) concluded that this class of medication was less effective at improving performance on tasks that place considerable demands on executive functions (such as working memory) than on tasks without such demands. They further stated that there is no "clear evidence of completely correcting cognitive deficits associated with ADHD" (p. 211). That those with ADHD may show cognitive deficits even after treatment with medications is consistent with evidence that medications can improve long-term functional outcomes, but have generally not resulted in normalization of functioning (Prasad et al., 2013; Shaw et al., 2012). For example, a recent review of the effects of medications on academic performance in youth with ADHD concluded that long-term treatment is associated with relatively small improvements in standardized achievement scores, and even less so with long-term improvements in school grades and grade retention (Langberg and Becker, 2012). Thus, alternate approaches are needed to improve cognitive functioning in ADHD.

Behavioral Approaches

There is little research examining the effects of behavioral interventions on working memory in ADHD. A small number of studies have examined whether provision of incentives may be helpful. In a study of 21 children with ADHD, Shiels et al. (2008) reported that incentives (feedback and points that could be exchanged for prizes) improved reverse digit span but not forward spatial span. Dovis and colleagues examined the impact of incentives on spatial working memory in 30 children with ADHD and 31 typically developing children (Dovis, Van der Oord, Wiers, and Prins, 2012). Results revealed that incentives significantly improved working memory. Even the strongest incentives (i.e., the largest amount of monetary reward or gaming), however, were unable to normalize performance in the clinical sample. Strand et al. reported a similar finding in their study of spatial working memory in 24 children with ADHD and 32 typically developing children (Strand et al., 2012). Provision of incentives (points exchanged for prizes) improved working memory in the children with ADHD, but the difference between the clinical and control groups was reduced by only about half. Performance was enhanced when incentives and treatment with methylphenidate were combined. The effects of other forms of behavioral interventions that have been developed for children having difficulty with working memory (e.g. Elliott, Gathercole, Alloway, Holmes, and Kirkwood, 2010) remain to be investigated in those with ADHD.

Cognitive Remediation of Working Memory in ADHD

Over the past decade investigators have begun to examine whether cognitive remediation of working memory may be beneficial for those with ADHD. Cognitive remediation typically involves the use of either paper-and-pencil or computerized cognitive tasks that are purported to help improve working memory (or other cognitive functions). Although there is considerable heterogeneity with respect to the specific methods used (e.g., nature of the tasks administered, duration of treatment), remediation generally requires a person to repeatedly complete one or more training tasks within a session as well as across multiple sessions that may span several weeks to months. Some of the approaches use adaptive training, that is, the difficulty level of the tasks used to "train" working memory varies trial-by-trial with the trainee's success and failure, becoming more difficult with success in an effort to encourage maximal engagement of working memory. A comprehensive review of cognitive remediation is beyond the scope of this chapter, and thus the reader is referred to a number of excellent sources on the topic for more information (Sohlberg and Mateer, 2001; Stuss, Winocur, and Robertson, 2010; Wilson, 2003).

We conducted a literature search using PubMed and the following search terms: ADHD, attention deficit, working memory, cognitive remediation, training, and therapy. This search yielded a total of 355 published articles. We then eliminated from consideration review papers, empirical studies that did not involving remediation of working memory, and studies that did not assess the effects of remediation in individuals with ADHD.

This yielded 12 studies that directly assessed the effects of remediation of working memory in ADHD. Table 1 provides information pertaining to each of these studies. Studies were reviewed for several salient outcome measures including improvements on the trained tasks, near transfer, far transfer, effects on ADHD symptoms, and maintenance of treatment gains.

Effects on Trained Tasks

Studies on remediation of working memory in ADHD have consistently shown improvements over time on the trained tasks. For example, Johnstone and colleagues used computer-based adaptive training tasks involving spatial working memory and impulse control in children with ADHD and typically developing children (Johnstone et al., 2012). At the end of the five week training period, both groups were performing the working memory task at a significantly higher level of difficulty than at the start of training. Children with ADHD were also found over time to show better performance, as well as time on task (thought to reflect motivation), on a spatial working memory training task with game-like elements such as a story line involving saving the world from evil robots (Prins, Dovis, Ponsioen, ten Brink, and van der Oord, 2011). A study using Cogmed (a computerized adaptive working memory training program) in adolescents with ADHD also showed considerable improvement on the trained tasks over the first few training days, with performance remaining relatively stable across the remainder of the training sessions (Gibson et al., 2011). This effect was observed regardless of whether training emphasized verbal or spatial working memory. As noted above, cognitive remediation programs typically require the investment of considerable time, often involving repeated sessions over multiple weeks. The Gibson et al. study raises the question of how much training is minimally needed to show desired outcomes, an issue that has not been addressed in published studies of ADHD.

Effects on Working Memory (Near Transfer)

Several studies have examined whether cognition remediation of working memory in ADHD shows near transfer effects. That is, do improvements seen on working memory tasks used in training generalize to similar, but untrained tasks also tapping working memory? Transfer to both spatial and verbal working memory tasks has been evaluated.

Transfer to non-trained spatial working memory tasks following Cogmed training, or an earlier iteration of Cogmed (Klingberg, Forssberg, and Westerberg, 2002), has been replicated by multiple studies. A significant treatment effect on both trained and non-trained spatial working memory tasks (span board) was found for children with ADHD as well as a group of healthy adults (Klingberg et al., 2002). These findings were replicated in a second study by the same research group using a larger sample of children with ADHD compared to a control group that completed a low-intensity control version of the training program (Klingberg et al., 2005).

In a study of adolescents with ADHD who completed Cogmed or a math training control program (Gray et al., 2012), working memory training resulted in greater improvements in performance on a measure of spatial span, but not on a measure of strategy use during spatial working memory.

Transfers effects have also been reported on non-trained verbal working memory tasks. Holmes et al. examined the effects of medication and Cogmed training on working memory in children with ADHD (Holmes et al., 2010). While both Cogmed and medication groups demonstrated gains in spatial working memory, only the Cogmed group showed improvements on verbal working memory transfer tasks. In a sample of children with attention problems or hyperactivity, Mezzacappa and Buckner (2010) observed improvements on both verbal (backward span) and spatial (Finger-Windows from the WRAML) working memory tasks following Cogmed training. Prins et al. also reported improved spatial working memory, as reflected in Corsi block tapping test performance, using a computerized training program (Prins et al., 2011).

Although this latter study did not find an effect of Cogmed training on block tapping span, the version of Cogmed employed differed from the published methods in several respects, as noted by the study authors. Green et al. reported greater improvement on the WISC-IV Working Memory Index in children with ADHD who completed the adaptive Cogmed training as compared to a group who completed training using a non-adaptive, low-level of difficulty version of the program (Green et al., 2012).

Two studies examined whether cognitive remediation of working memory showed generalization to working memory abilities as reflected in the everyday life of children with ADHD. One study found improvement on the Working Memory scale of the parent but not the teacher report BRIEF (Beck, Hanson, Puffenberger, Benninger, and Benninger, 2010), though the finding was not replicated in the waitlist control group after they also completed Cogmed. The second study used the Working Memory Rating Scale (Alloway et al., 2009) and found no effect of training on teacher ratings (Gray et al., 2012).

Table 1. Characteristics of Reviewed Studies

Authors	Participant characteristics	Age Range (years)	Study design	Training program (n)	Control Training (n)	Near transfer	Far transfer	ADHD Symptom Assessment	Follow up
Beck et al. (2010)	ADHD	7-17	Randomized, controlled	Cogmed (27)	Waitlist (25)	BRIEF Working Memory scale (Parent and Teacher)	BRIEF (Parent and Teacher)	CPRS-R (Parent and Teacher)	4 mo.
Gibson et al. (2011)	ADHD	11-16	Randomized	Modified Cogmed (24 verbal, 23 spatial)	-	Verbal & spatial IFR tasks	-	DuPaul ADHD Scale (Parent and Teacher)	Post-test only
Gray et al. (2012)	ADHD with LD	12-17	Randomized, controlled	Cogmed (32)	Math Training Program (20)	CANTAB SSP, DSB, WMRS	D2 Test of Attention, WRAT-4	IOWA Conners scale (Parent and Teacher), SWAN (Parent and Teacher)	3 wks
Green et al. (2012)	ADHD	7-14	Randomized, controlled, double-blind	Cogmed (12)	Non-Adaptive Cogmed (14)	WISC-IV WM Index	-	CPRS-R and CTRS-R (Parent and Teacher), RAST (objective rater)	DNR
Holmes et al. (2010)	ADHD	8-11	Within-Subject	Cogmed (25)	-	AWMA verbal & spatial working memory tasks	AWMA verbal & spatial short-term memory tasks, WASI	-	6 mo.
Johnstone et al., (2010)	ADHD	7-12	Randomized, controlled double-blind	Adaptive WM and IC (15)	Low intensity Training (14)	-	Go/No-go task	CPRS-R, CBCL parent, Purpose-designed ADHD sx questionnaire (Parent and Other Informant)	Post-test only
Johnstone et al. (2012)	ADHD, Healthy control	7-13	Randomized, controlled	Adaptive WM and IC: ADHD (40) Control (43)	Waitlist: ADHD (20) Control (25)	FTM	Oddball Task, Flanker Task, Go/No-go Task, Counting Span	CPRS-R (Parent), Purpose designed BRS (Parent and Other Informant)	6 wks.

Authors	Participant characteristics	Age Range (years)	Study design	Training program (n)	Control Training (n)	Near transfer	Far transfer	ADHD Symptom Assessment	Follow up
Kingberg et al. (2002)	Experiment 1: ADHD	7-15	Randomized, controlled, double-blind	Adaptive computerized WM (7)	Low intensity version of WM (7)	Spatial WM Span Board task	Stroop, Raven's Matrices, Choice reaction time task	Head Movements	Post-test only
	Experiment 2: Healthy Adult	20-29	Within-Subject	Adaptive computerized WM (4)	-	Spatial WM Span Board task	Stroop, Raven's Matrices, Choice reaction time task	-	Post-test only
Klingberg et al. (2005)	ADHD	7-12	Randomized, controlled, double-blind	Cogmed (20)	Low intensity version of WM (24)	Span Board, Digit-Span	Stroop, Raven's Matrices	CPRS & CTRS (Parent and Teacher), Head Movements	3 mo.
Mezzacappa et al. (2010)	ADHD	8-10	Within-Subject	Cogmed (9)	-	Digit Span, WRAML FingerWindows	-	ADHD-RS-IV (Teacher)	Post-test only
Oord et al. (2012)	ADHD	8-12	Randomized, controlled	Braingame Brian (18)	Waitlist (22)	-	-	DBDRS (Parent and Teacher), BRIEF (Parent)	9 wks.
Prins et al. (2011)	ADHD	7-12	Randomized, controlled	Adaptive computerized WM Training with game elements (27)	WM Training without game elements (24)	Corsi Block Tapping	-	-	Post-test only

Note. - = not included in study; ADHD-RS-IV = ADHD Rating Scale IV; AWMA = Automated Working Memory Assessment; BRIEF = Behavior Rating Inventory of Executive Function; BRS = Behavior Rating Scale; CANTAB SSP = Cambridge Neuropsychological testing Automated Battery short-term visual-spatial storage; CBCL = Child Behavior Checklist; CBCT = Corsi Block-Tapping Test; CPRS-R = Conner's Parent Rating Scale-Revised; CTRS-R = Conner's Teacher Rating Scale-Revised; DBDRS = Disruptive Behavior Disorder Rating Scale; DSB = Digit Span Backward; IC = Inhibitory Control; IFR = Immediate Free Recall; RAST = Restricted Academic Situations Task; SWAN = Strengths and Weakness of ADHD-symptoms and Normal-behavior Scale; WASI = Wechsler Abbreviated Scales of Intelligence, WISC-IV = Wechsler Intelligence Scale for Children-Fourth Edition; WOND = Wechsler Objective Number Dimensions; WORD = Wechsler Objective Reading Dimensions; WMRS = Working Memory Rating Scale; WRAT-4 = Wide Range Achievement Test; WRAML = Wide Range Assessment of Memory and Learning.

Effects on Other Cognitive Functions (Far Transfer)

A number of studies of working memory training in ADHD have reported evidence of generalization to cognitive functions other than working memory. Improvements have been reported for response inhibition and non-verbal reasoning/problem-solving (Holmes et al., 2010; Klingberg et al., 2005; Klingberg et al., 2002), as well as verbal and spatial short-term memory (Holmes et al., 2010). In two studies involving children with ADHD, Johnstone and colleagues observed generalization of the effects of a combined working memory and impulse control training program (Johnstone et al., 2012; Johnstone, Roodenrys, Phillips, Watt, and Mantz, 2010). Specifically, training was associated with reduced interference from distracting stimuli (flanker task), better attention (oddball task), as well as reduced response time variability and both omission or commission errors (go/no-go task). Taken together, the findings from Johnstone et al. suggest that children with ADHD show better impulse control and attention after adaptive working memory and inhibitory control training; though it is unclear to what extent the benefits may be attributed to the working memory component of the training per se as there was no comparison group that completed only working memory training .

With respect to everyday behavior, improvement with training in children with ADHD was observed by parents, but not teachers, on the BRIEF scales of Initiate and Plan/Organize (Beck et al., 2010). Similarly, a study involving training of several executive functions (spatial working memory, inhibition and cognitive flexibility) found that parents of children with ADHD reported improvement on the BRIEF Metacognition Index (reflecting initiation, working memory, monitoring, planning and organization), but not Shift scale (ability to think flexibly) compared to those in a waitlist control (Oord, Ponsioen, Geurts, Brink, and Prins, in press).

Despite these promising findings, there has been only limited research addressing far transfer in ADHD and negative results have also been reported. For example, no significant effect of training was found for measures of auditory-verbal short term memory (digit span forward), academic skills (Gray et al., 2012), or intellectual functioning (Holmes et al., 2010). Furthermore, generalization to functioning in everyday life (cognition, academic, adaptive) has received scant empirical attention.

Transfer to ADHD Symptoms

Along with cognition, studies of ADHD have typically also assessed whether working memory training improves core symptoms of the disorder. A variety of symptom measures have been employed, though these have primarily consisted of observational rating scales such as the Connors Rating Scales (CPRS-R; Conners, 1997).

Studies using Cogmed in children with ADHD have reported improved parent ratings of inattention (Beck et al., 2010; Klingberg et al., 2005) and CPRS-R ADHD Index (Beck et al., 2010) scores, but not hyperactivity or oppositional behavior (Beck, et al., 2010). In a study of adolescents with ADHD, those who showed the most improvement on the Cogmed training tasks (administered at school) were rated by parents as having the greatest reduction in inattention and hyperactivity in the home (Gray et al., 2012). In contrast, one study found improved ADHD Index scores regardless of whether the child was in the training or control group (Green et al., 2012).

Changes in ADHD symptoms have also been reported following training involving different paradigms. Parent rated inattention and hyperactivity/impulsivity improved in a study using modified versions of Cogmed tasks coupled with additional inhibitory control training tasks (Gibson et al., 2011). Children in the working memory and inhibition training group were rated by both parents and another adult informant, relative to a waitlist control group, as being more improved on symptom rating scales (Johnstone et al., 2012). Significant reductions in inattention and hyperactivity were noted in an earlier study by the same group (Johnstone et al., 2010). Importantly, high intensity training was associated with greater reduction of ADHD symptoms than low intensity training. Children with ADHD who completed a program that trained working memory as well as other executive functions showed greater reductions in inattention and hyperactivity/ impulsivity compared to those in a waitlist control group (Oord et al., in press).

It is important to note that in contrast to parent reports, teacher ratings have generally not shown significant improvements in ADHD symptoms with training (Gray et al., 2012; Green et al., 2012; Oord et al., in press). At present, there are only a few exceptions to this finding. In a small sample pilot study of school-based implementation of Cogmed, teacher ratings of ADHD symptoms (using the ADHD-RS-IV) improved by 26% overall (Mezzacappa and Buckner, 2010). Another study observed improved ADHD symptoms as rated by both parents and teachers following working memory training (Gibson et al., 2011). Nevertheless, factors that are impacting the differential observations by parents and teachers clearly require further investigation.

In addition to rating scales, a few studies have examined ADHD symptoms using other forms of measurement. Klingberg et al. reported that working memory training reduced head movements, and that this reduction was associated with greater improvements in working memory (Klingberg, et al., 2002). A recent study observed that working memory training resulted in better performance on a simulated classroom setting task that allowed for observation and quantification of ADHD-associated behaviors (Green et al., 2012). In particular, the greatest improvement was seen in the ability of children to focus on the task at hand, suggesting amelioration of inattention.

Maintenance of Treatment Gains

Most of the studies of ADHD that have examined whether the effects of working memory training are maintained over time have reported relative stability of gains. The longest follow-up period, however, was six months, with most studies re-assessing participants no more than three months after the end of training and considerable variability in the time between the end of training and follow-up assessment.

In terms of ADHD symptoms, several studies have reported that improvements seen by the end of training continue to be seen at the time of follow-up evaluations. This has been found for a number of different follow-ups periods including three weeks (Gray et al., 2012), six weeks (Johnstone et al., 2012), nine weeks (Oord et al., in press), three months (Klingberg et al., 2005), and four months (Beck et al., 2010).

Some maintenance of gains on near and far transfer measures has also been observed. Parent rated executive functioning on the BRIEF has been reported to show benefits of training at nine weeks and four months (Beck et al., 2010; Oord et al., in press). Sustained improvements on auditory-verbal and visual-spatial working memory tasks have been reported at a three week follow-up (Gray et al., 2012). Another study reported that gains in response inhibition, problem-solving/complex reasoning, and spatial working memory were maintained at the three month assessment (Klingberg et al., 2005). The study with the longest follow-up period (six months) observed continued benefits of training for spatial working memory, verbal working memory and spatial short-term memory, although verbal short term memory declined (Holmes et al., 2010).

METHODOLOGICAL CONSIDERATIONS

Participant Characteristics

In general, studies of ADHD have involved relatively modest sample sizes. At the low end, only four participants were included in a study of adult ADHD (Klingberg et al., 2002), while a pilot study included only nine children with the disorder (Mezzacappa and Buckner, 2010). Most study samples, however, have ranged from 12 to 40 patients in the working memory training condition (see Table 1).

Diagnostic characteristics have varied somewhat, with a small number of studies restricting participation to individuals with a specific ADHD subtype (Holmes et al., 2010; Johnstone et al., 2010). Efforts to limit diagnostic comorbidities among participants with ADHD have been inconsistent. Some studies excluded comorbid psychiatric disorders (Green et al., 2012; Johnstone et al., 2012; Johnstone et al., 2010) or specified the exclusion of individuals with head injury, seizure, or hearing and vision problems (Johnstone et al., 2012; Johnstone et al., 2010). One study recruited students selected for comorbid ADHD and learning disorder, though further evaluation revealed that a subset did not have this comorbidity (Gray et al., 2012). Another excluded children with IQ below 80 or who were taking atomoxetine, but noted that 40% of their final sample also had oppositional defiant disorder (Oord et al., in press). Yet another study included children with ADHD and comorbid conduct or oppositional defiant disorder, anxiety or mood disorders, with 63% of participants having at least one and as many as three comorbid psychiatric disorders (Beck et al., 2010). Some studies were unclear as to whether they assessed for psychiatric comorbidities in their samples of children with ADHD (e.g. Beck et al., 2010; Klingberg et al., 2002). The impact of such comorbidities on outcome following working memory training has not been systematically investigated.

Another important consideration that has been only inconsistently addressed is whether individuals recruited for studies of working memory training programs actually have a deficit in this cognitive process; as established using standardized measures rather than the training tasks (e.g., Beck et al., 2010). Although there is considerable evidence for the presence of working memory deficit in ADHD (see above), there is also substantial variability in neuropsychological test performance within ADHD samples (Willcutt et al., 2005). Furthermore, a recent study found significantly better cognitive and behavioral response to methylphenidate in children with ADHD having pre-treatment impairments in working memory and self-regulation than those whose were relatively intact (Hale et al., 2011). Such findings raise the possibility that heterogeneous outcomes may be related in part to the pre-treatment cognitive characteristics of study participants, and are consistent with concerns regarding the impact of individual difference factors on cognitive training outcomes in ADHD (Shah, Buschkuehl, and Jonides, 2012).

Study Design

Studies have varied considerably in their use and type of control groups. Some investigations have employed a non-adaptive/low intensity version of Cogmed to contrast with the effects of the standard Cogmed training program (Green et al., 2012; Klingberg et al., 2005; Klingberg et al., 2002). Johnstone et al. (Johnstone et al., 2010) similarly used an active control group that completed a lower intensity version of their training program. Other work has used a computerized math training program as an active comparison group (Gray et al., 2012). One pilot study did not employ any form of control group (Mezzacappa and Buckner, 2010). Holmes et al. (2010) utilized a within-group nonrandomized design in which they assessed performance on medication alone and then again after Cogmed training (Holmes et al., 2010). Two studies employed waitlist control groups that engaged in treatment once the experimental group completed treatment (Beck et al., 2010; Oord, et al., in press). This allowed for a larger overall training sample size, but cohort effects likely contributed to inconsistent findings between two studies despite equivalent treatment protocols.

Nature of the Training Program

The majority of the studies of cognitive remediation in ADHD have used the Cogmed working memory training program, which is computer-based and involves adaptive training using a variety of tasks that require working memory (see Table 1). Cogmed is typically completed by the client in their home environment, which in studies using other remediation tools has been shown to also yield significant benefit (Boman, Lindstedt, Hemmingsson, and Bartfai, 2004). Other studies have used adaptations of Cogmed with the inclusion of additional working memory training tasks (Gibson et al., 2011), working memory training with additional game elements (Prins et al., 2011), or training programs designed to remediate working memory as well as other executive functions (Johnstone et al., 2012; Oord et al., in press).

Variability in the implementation of working memory training programs carried out in the home environment may also have contributed to discrepant findings. Most studies trained parents of participants on use of the software training program at home. This raises the possibility of potential confounds such as level of parental involvement in their child's training exercises. For example, a greater degree of parental assistance and encouragement to complete training tasks, as well as more parental attention in a general sense, may be especially important for children and adolescents with ADHD. This could inflate not only training versus control group differences on the trained tasks, but additionally influence parent ratings of change with training. Evidence of such an effect may be found in a placebo-controlled, randomized, double-blind study in which both adaptive and non-adaptive (control) groups demonstrated improvements on parental ratings of ADHD symptoms (Green et al., 2012). Furthermore, adding child-friendly game elements to working memory training resulted in greater motivation, more time spent training, better performance on training tasks, as well as better working memory post-training than a program without such game elements (Prins et al., 2011). Such findings suggest that motivational factors are likely important in improving working memory, consistent with the findings from behavioral studies noted above.

Selection of Outcome Measures

Another likely important contributor to the heterogeneity of findings is the variability in outcome measures used to assess near and far transfer. As seen in table 1, a wide variety of instruments have been employed to assess working memory, other cognitive functions, and ADHD symptomatology. Furthermore, there has been limited attention paid to assessing changes in everyday life, other than ADHD related symptoms. For example, it would be important to determine whether improvements in working memory seen in training generalize to measures such as those assessing academic and occupational performance, subjective ratings of cognitive abilities, and other relevant problems common to ADHD (e.g., driving ability in adults with the disorder).

Follow Up Period

There is a significant dearth in the measurement of the enduring effects of working memory training in ADHD. This is particularly unfortunate as cognitive remediation could potentially be associated with improvement that may not be apparent over the short term, such as academic performance. The longest follow up period examined thus far has been 6 months (Holmes et al., 2010; Johnstone et al., 2012). Several studies have employed shorter follow up periods, such as three weeks (Gray et al., 2012), 9 weeks (Oord et al., in press), 3 months (Klingberg, et al., 2005), and 4 months (Beck et al., 2010). Other studies did not include assessment beyond the immediate post-treatment period (Green et al., 2012; Johnstone et al., 2010; Klingberg et al., 2002; Mezzacappa and Buckner, 2010). Thus, it remains unclear to what extent any treatment gains are maintained and what if any factors may be associated with better or worse long term outcome following working memory training.

Expectancy Effects

The majority of studies reviewed utilized informant-based rating scales to measure treatment effects on ADHD symptoms. While some of these studies involved parents and teachers blind to the participant's group assignment (training vs. control), some did not blind informants or blinded one but not the other informant (Beck et al., 2010; Gray et al., 2012; Mezzacappa and Buckner, 2010), or did not indicate if informants were blinded (Johnstone et al., 2012). Subjective ratings scales may be sensitive to expectancy effects. As most of the studies utilized working memory training that was conducted in the home or school setting, with regular instruction by parents or teachers, knowledge of group placement by informants may have impacted the delivery of the training protocol and/or ratings of outcome. For example, if a parent was aware that their child was receiving the treatment rather than control condition, their expectations for improvement could have affected their child's commitment to the training exercises. This effect may have contributed to the divergent ratings of ADHD symptoms and executive functions by parents (not blind) as compared to teachers (blinded) in the study by Beck et al. (Beck et al., 2010).

Effects of Concurrent Medications

Participants in several of the studies remained on their prescribed medication for ADHD throughout the study (Beck et al., 2010; Gray et al., 2012; Green et al., 2012; Johnstone et al., 2010; Johnstone et al., 2012, Klingberg et al, 2002; Mezzacappa et al., 2010). This renders it difficult to determine to what extent treatment gains were the result of working memory training per se versus an interaction between training and medication. The latter possibility was supported in a study that examined a subsample of medicated children who underwent training and concluded that there may be an additive effect of methylphenidate and training (Oord, et al., in press). Another study noted that including medication as a covariate in analyses did not account for the effects of training on objective measures of working memory or a blind rater measure of observed ADHD behavior, but may have contributed to parental rating of ADHD behavior (Green et al., 2012). Further evidence that working memory training has benefits that are independent from medication in those with ADHD comes from work that only included participants that were free from stimulant and other psychoactive medications (Klingberg et al., 2005).

MECHANISMS OF CHANGE

There is extensive literature addressing the neural basis of working memory in humans (D'Esposito et al., 1995; Wager and Smith, 2003). Studies employing functional magnetic resonance imaging (fMRI) have revealed widespread abnormalities in neural circuitry subserving working memory in children and adults with ADHD (Bayerl et al., 2010; Fassbender et al., 2011; Massat et al., 2012; Mills et al., 2012; Valera, Faraone, Biederman, Poldrack, and Seidman, 2005; Wolf et al., 2009). These have implicated a distributed circuitry involving several brain regions including the frontal and parietal lobes, thalamus, basal ganglia, and cerebellum.

Evidence suggests that some medications, particularly dopaminergic ones, may improve working memory circuitry in ADHD. For example, methylphenidate or a dextroamphetamine/amphetamine combination improved fMRI functional connectivity of the circuitry in a randomized, double-blind, placebo-controlled study of children and adolescents with the disorder (Wong and Stevens, 2012). Working memory training has also been reported to result in fMRI changes in this circuitry in healthy adults (Brehmer et al., 2011;

Olesen, Westerberg, and Klingberg, 2004; Westerberg and Klingberg, 2007), but similar studies have yet to be reported in ADHD.

The mechanism by which changes in working memory or far transfer are achieved is poorly understood at present. A commonly raised possibility is that cognitive remediation programs promote changes in neural plasticity, which may be mediated by the neurotransmitter dopamine (Klingberg, 2010; Soderqvist et al., 2012). Consistent with this hypothesis, the degree of improvement in working memory following working memory training has been associated with polymorphisms of certain dopamine genes (Bellander et al., 2011; Brehmer et al., 2009; Soderqvist et al., 2012). Furthermore, two positron emission tomography studies have reported that changes with working memory training are associated with alterations in the dopamine system (Backman et al., 2011; McNab et al., 2009). Such findings are in line with evidence that dopamine plays an important role in neural plasticity in animals (Gonzalez-Burgos, Kroener, Seamans, Lewis, and Barrionuevo, 2005; Young and Yang, 2005). Nonetheless, it should be noted that a recent review of neuroimaging studies involving working memory training concluded that there is no clear evidence for any specific mechanism involved in training or transfer of training effects (Buschkuehl, Jaeggi, and Jonides, 2012). Multidisciplinary research investigating potential mechanisms of change with working memory training in ADHD will therefore be important.

CONCLUSION

Cognitive remediation of working memory has been receiving increasing empirical attention in ADHD and other populations. Moreover, training of working memory and other cognitive skills such as episodic memory has been the subject of considerable media attention. The use of training programs, nonetheless, remains controversial.

The evidence base specifically addressing the effectiveness of working memory training in ADHD is growing. There is considerable variability in study methodologies that limits cross-study comparisons, however, including use of a variety of different training paradigms, Cogmed having received the most empirical attention in the ADHD literature. With these caveats, the available literature points to at least some beneficial effect of cognition remediation of working memory in ADHD, particularly when using an adaptive, high intensity training approach. Findings appear to be most consistent for ADHD symptoms (though mostly as rated by parents rather than by teachers) and near transfer tasks. Somewhat greater variability was noted for far transfer tasks, and there is as yet little information on whether training can benefit everyday functioning (e.g., school or job performance). It remains unclear whether gains made following working memory training are maintained over the long-term, as few studies have followed study participants beyond three months after the end of training.

In sum, cognitive remediation of working memory shows promise as novel tool in the armamentum of those treating ADHD. Further research will be important to determine the generalizability of training effects beyond working memory, the longevity of treatment gains, as well as to identify individual differences and other factors that could help predict good and poor response to training.

REFERENCES

Agarwal, R., Goldenberg, M., Perry, R., and IsHak, W. W. (2012). The quality of life of adults with attention deficit hyperactivity disorder: A systematic review. *Innovations in Clinical Neuroscience,* 9(5-6), 10-21.

Alderson, R. M., Rapport, M. D., Hudec, K. L., Sarver, D. E., and Kofler, M. J. (2010). Competing core processes in attention-deficit/hyperactivity disorder (ADHD): Do working memory deficiencies underlie behavioral inhibition deficits? *Journal of Abnormal Child Psychology,* 38(4), 497-507.

Alloway, T. P. (2011). A comparison of working memory profiles in children with ADHD and DCD. *Child Neuropsychology,* 17(5), 483-494.

Alloway, T. P., Gathercole, S. E., and Elliott, J. (2010). Examining the link between working memory behaviour and academic attainment in children with ADHD. *Developmental Medicine and Child Neurology,* 52(7), 632-636.

Alloway, T. P., Gathercole, S. E., Holmes, J., Place, M., Elliott, J. G., and Hilton, K. (2009). The diagnostic utility of behavioral checklists in identifying children with ADHD and children with working memory deficits. *Child Psychiatry and Human Development,* 40(3), 353-366.

Association, A. P. (2000). Diagnostic and Statistical Manual of Mental Disorders (revised, 4th ed.). Washington, DC: American Psychiatric Press.

Backman, L., Nyberg, L., Soveri, A., Johansson, J., Andersson, M., Dahlin, E., . . . Rinne, J. O. (2011). Effects of working-memory training on striatal dopamine release. *Science,* 333(6043), 718.

Baddeley, A. (1992). Working memory. *Science,* 255(5044), 556-559.

Baddeley, A. (1996). The fractionation of working memory. *Proceedings of the National Academy of Sciences of the United States of America,* 93(24), 13468-13472.

Baddeley, A. (2010). Working memory. *Current Biology,* 20(4), R136-140.

Barkley, R. A. (1997a). Attention-deficit/hyperactivity disorder, self-regulation, and time: Toward a more comprehensive theory. *Journal of Developmental and Behavioral Pediatrics,* 18(4), 271-279.

Barkley, R. A. (1997b). Behavioral inhibition, sustained attention, and executive functions: Constructing a unifying theory of ADHD. *Psychological Bulletin,* 121(1), 65-94.

Barkley, R. A., Fischer, M., Smallish, L., and Fletcher, K. (2002). The persistence of attention-deficit/hyperactivity disorder into young adulthood as a function of reporting source and definition of disorder. *Journal of Abnormal Psychology,* 111(2), 279-289.

Barkley, R. A., Murphy, K. R., Dupaul, G. I., and Bush, T. (2002). Driving in young adults with attention deficit hyperactivity disorder: knowledge, performance, adverse outcomes, and the role of executive functioning. *Journal of the International Neuropsychological Society,* 8(5), 655-672.

Bayerl, M., Dielentheis, T. F., Vucurevic, G., Gesierich, T., Vogel, F., Fehr, C., . . . Konrad, A. (2010). Disturbed brain activation during a working memory task in drug-naive adult patients with ADHD. *Neuroreport,* 21(6), 442-446.

Beck, S. J., Hanson, C. A., Puffenberger, S. S., Benninger, K. L., and Benninger, W. B. (2010). A controlled trial of working memory training for children and adolescents with ADHD. *Journal of Clinical Child and Adolescent Psychology,* 39(6), 825-836.

Bedard, A.-C., Jain, U., Johnson, S. H., and Tannock, R. (2007). Effects of methylphenidate on working memory components: Influence of measurement. *Journal of Child Psychology and Psychiatry and Allied Disciplines,* 48(9), 872-880.

Bedard, A. C., and Tannock, R. (2008). Anxiety, methylphenidate response, and working memory in children with ADHD. *Journal of Attention Disorders,* 11(5), 546-557.

Bellander, M., Brehmer, Y., Westerberg, H., Karlsson, S., Furth, D., Bergman, O., . . . Backman, L. (2011). Preliminary evidence that allelic variation in the LMX1A gene influences training-related working memory improvement. *Neuropsychologia,* 49(7), 1938-1942.

Berlin, L., Bohlin, G., Nyberg, L., and Janols, L. O. (2004). How well do measures of inhibition and other executive functions discriminate between children with ADHD and controls? *Child Neuropsychology,* 10(1), 1-13.

Bidwell, L. C., McClernon, F. J., and Kollins, S. H. (2011). Cognitive enhancers for the treatment of ADHD. *Pharmacology, Biochemistry and Behavior,* 99(2), 262-274.

Blum, N. J., Jawad, A. F., Clarke, A. T., and Power, T. J. (2011). Effect of osmotic-release oral system methylphenidate on different domains of attention and executive functioning in children with attention-deficit-hyperactivity disorder. *Developmental Medicine and Child Neurology,* 53(9), 843-849.

Boman, I. L., Lindstedt, M., Hemmingsson, H., and Bartfai, A. (2004). Cognitive training in home environment. *Brain Injury,* 18(10), 985-995.

Brehmer, Y., Rieckmann, A., Bellander, M., Westerberg, H., Fischer, H., and Backman, L. (2011). Neural correlates of training-related working-memory gains in old age. *Neuroimage,* 58(4), 1110-1120.

Brehmer, Y., Westerberg, H., Bellander, M., Furth, D., Karlsson, S., and Backman, L. (2009). Working memory plasticity modulated by dopamine transporter genotype. *Neuroscience Letters,* 467(2), 117-120.

Brocki, K. C., Randall, K. D., Bohlin, G., and Kerns, K. A. (2008). Working memory in school-aged children with attention-deficit/hyperactivity disorder combined type: Are deficits modality specific and are they independent of impaired inhibitory control? *Journal of Clinical and Experimental Neuropsychology,* 30(7), 749-759.

Buschkuehl, M., Jaeggi, S. M., and Jonides, J. (2012). Neuronal effects following working memory training. *Developmental Cognitive Neuroscience, 2 Suppl 1,* S167-179.

Clark, C., Prior, M., and Kinsella, G. (2002). The relationship between executive function abilities, adaptive behaviour, and academic achievement in children with externalising behaviour problems. *Journal of Child Psychology and Psychiatry and Allied Disciplines,* 43(6), 785-796.

Conners, K. (1997). Conners' Rating Scales - Revised: Technical Manual. Tonawanda, NY: Multi-Health Systems Inc.

D'Esposito, M., Detre, J. A., Alsop, D. C., Shin, R. K., Atlas, S., and Grossman, M. (1995). The neural basis of the central executive system of working memory. *Nature,* 378(6554), 279-281.

Danckaerts, M., Sonuga-Barke, E. J., Banaschewski, T., Buitelaar, J., Dopfner, M., Hollis, C., . . . Coghill, D. (2010). The quality of life of children with attention deficit/hyperactivity disorder: A systematic review. *European Child and Adolescent Psychiatry,* 19(2), 83-105.

de Graaf, R., Kessler, R. C., Fayyad, J., ten Have, M., Alonso, J., Angermeyer, M., . . . Posada-Villa, J. (2008). The prevalence and effects of adult attention-deficit/hyperactivity disorder (ADHD) on the performance of workers: Results from the WHO World Mental Health Survey Initiative. *Occupational and Environmental Medicine,* 65, 835 - 842.

Dovis, S., Van der Oord, S., Wiers, R. W., and Prins, P. J. (2012). Can motivation normalize working memory and task persistence in children with attention-deficit/hyperactivity disorder? The effects of money and computer-gaming. *Journal of Abnormal Child Psychology,* 40(5), 669-681.

Dowson, J. H., McLean, A., Bazanis, E., Toone, B., Young, S., Robbins, T. W., and Sahakian, B. J. (2004). Impaired spatial working memory in adults with attention-deficit/hyperactivity disorder: Comparisons with performance in adults with borderline personality disorder and in control subjects. *Acta Psychiatrica Scandinavica,* 110(1), 45-54.

Elliott, J. G., Gathercole, S. E., Alloway, T. P., Holmes, J., and Kirkwood, H. (2010). Evaluation of a classroom-based intervention to help overcome working memory difficulties and improve long-term academic achievement. *Journal of Cognitive Education and Psychology*, 9(3), 227-250.

Fassbender, C., Schweitzer, J. B., Cortes, C. R., Tagamets, M. A., Windsor, T. A., Reeves, G. M., and Gullapalli, R. (2011). Working memory in attention deficit/hyperactivity disorder is characterized by a lack of specialization of brain function. *PLoS ONE,* 6(11), e27240.

Fayyad, J., De Graaf, R., Kessler, R., Alonso, J., Angermeyer, M., Demyttenaere, K., . . . Jin, R. (2007). Cross-national prevalence and correlates of adult attention-deficit hyperactivity disorder. *British Journal of Psychiatry,* 190, 402 - 409.

Finke, K., Schwarzkopf, W., Muller, U., Frodl, T., Muller, H. J., Schneider, W. X., . . . Hennig-Fast, K. (2011). Disentangling the adult attention-deficit hyperactivity disorder endophenotype: Parametric measurement of attention. *Journal of Abnormal Psychology,* 120(4), 890-901.

Gathercole, S. E., Alloway, T. P., Willis, C., and Adams, A.-M. (2006). Working memory in children with reading disabilities. *Journal of Experimental Child Psychology,* 93(3), 265-281.

Gau, S. S., and Shang, C. Y. (2010). Improvement of executive functions in boys with attention deficit hyperactivity disorder: An open-label follow-up study with once-daily atomoxetine. *International Journal of Neuropsychopharmacology,* 13(2), 243-256.

Gibson, B. S., Gondoli, D. M., Johnson, A. C., Steeger, C. M., Dobrzenski, B. A., and Morrissey, R. A. (2011). Component analysis of verbal versus spatial working memory training in adolescents with ADHD: A randomized, controlled trial. *Child Neuropsychology,* 17(6), 546-563.

Gilotty, L., Kenworthy, L., Sirian, L., Black, D. O., and Wagner, A. E. (2002). Adaptive skills and executive function in autism spectrum disorders. *Child Neuropsychology,* 8(4), 241-248.

Gioia, G. A., Isquith, P. K., Guy, S. C., and Kenworthy, L. (2000). *BRIEF: Behavior rating inventory of executive function.* Lutz, Florida: Psychological Assessment Resources, Inc.

Goldman-Rakic, P. S. (1992). Working memory and the mind. *Scientific American,* 267, 110-117.

Gonzalez-Burgos, G., Kroener, S., Seamans, J. K., Lewis, D. A., and Barrionuevo, G. (2005). Dopaminergic modulation of short-term synaptic plasticity in fast-spiking interneurons of primate dorsolateral prefrontal cortex. *Journal of Neurophysiology,* 94(6), 4168-4177.

Gray, S. A., Chaban, P., Martinussen, R., Goldberg, R., Gotlieb, H., Kronitz, R., . . . Tannock, R. (2012). Effects of a computerized working memory training program on working memory, attention, and academics in adolescents with severe LD and comorbid ADHD: A randomized controlled trial. *Journal of Child Psychology and Psychiatry,* 53(12), 1277-1284.

Green, C. T., Long, D. L., Green, D., Iosif, A. M., Dixon, J. F., Miller, M. R., . . . Schweitzer, J. B. (2012). Will working memory training generalize to improve off-task behavior in children with attention-deficit/hyperactivity disorder? *Neurotherapeutics,* 9(3), 639-648.

Gropper, R. J., and Tannock, R. (2009). A pilot study of working memory and academic achievement in college students with ADHD. *Journal of Attention Disorders,* 12(6), 574-581.

Haberlandt, K. (1997). Cognitive psychology. Boston: Allyn and Bacon.

Hale, J. B., Reddy, L. A., Semrud-Clikeman, M., Hain, L. A., Whitaker, J., Morley, J., . . . Jones, N. (2011). Executive impairment determines ADHD medication response: Implications for academic achievement. *Journal of Learning Disabilities,* 44(2), 196-212.

Hinshaw, S. P. (2002). Preadolescent girls with attention-deficit/hyperactivity disorder: I. Background characteristics, comorbidity, cognitive and social functioning, and parenting practices. *Journal of Consulting and Clinical Psychology,* 70(5), 1086-1098.

Holmes, J., Gathercole, S. E., Place, M., Dunning, D. L., Hilton, K. A., and Elliot, J. G. (2010). Working memory deficits can be overcome: Impacts of training and medication on working memory in children with ADHD. *Applied Cognitive Psychology,* 24(6), 827-836.

Jarratt, K. P., Riccio, C. A., and Siekierski, B. M. (2005). Assessment of attention deficit hyperactivity disorder (ADHD) using the BASC and BRIEF. *Applied Neuropsychology,* 12(2), 83-93.

Johnstone, S. J., Roodenrys, S., Blackman, R., Johnston, E., Loveday, K., Mantz, S., and Barratt, M. F. (2012). Neurocognitive training for children with and without AD/HD. *Attention Deficit Hyperactivity Disorder,* 4(1), 11-23.

Johnstone, S. J., Roodenrys, S., Phillips, E., Watt, A. J., and Mantz, S. (2010). A pilot study of combined working memory and inhibition training for children with AD/HD. *Attention Deficit Hyperactivity Disorder,* 2(1), 31-42.

Jonsdottir, S., Bouma, A., Sergeant, J. A., and Scherder, E. J. (2005). The impact of specific language impairment on working memory in children with ADHD combined subtype. *Archives of Clinical Neuropsychology,* 20(4), 443-456.

Kasper, L. J., Alderson, R. M., and Hudec, K. L. (2012). Moderators of working memory deficits in children with attention-deficit/hyperactivity disorder (ADHD): A meta-analytic review. *Clinical Psychology Review,* 32(7), 605-617.

Katz, L. J., Brown, F. C., Roth, R. M., and Beers, S. R. (2011). Processing speed and working memory performance in those with both ADHD and a reading disorder compared with those with ADHD alone. *Archives of Clinical Neuropsychology,* 26(5), 425-433.

Kessler, R. C., Adler, L., Barkley, R., Biederman, J., Conners, C. K., Demler, O., . . . Zaslavsky, A. M. (2006). The prevalence and correlates of adult ADHD in the United

States: Results from the National Comorbidity Survey Replication. *American Journal of Psychiatry,* 163, 716-723.

Kessler, R. C., Lane, M., Stang, P. E., and Van Brunt, D. L. (2009). The prevalence and workplace costs of adult attention deficit hyperactivity disorder in a large manufacturing firm. *Psychological Medicine,* 39(1), 137-147.

Klein, R. G., Mannuzza, S., Olazagasti, M. A., Roizen, E., Hutchison, J. A., Lashua, E. C., and Castellanos, F. X. (2012). Clinical and functional outcome of childhood attention-deficit/hyperactivity disorder 33 years later. Archives of General Psychiatry, 69(12), 1295-1303.

Klingberg, T. (2010). Training and plasticity of working memory. *Trends in Cognitive Science,* 14(7), 317-324.

Klingberg, T., Fernell, E., Olesen, P. J., Johnson, M., Gustafsson, P., Dahlstrom, K., . . . Westerberg, H. (2005). Computerized training of working memory in children with ADHD: A randomized, controlled trial. *Journal of the American Academy of Child and Adolescent Psychiatry,* 44(2), 177-186.

Klingberg, T., Forssberg, H., and Westerberg, H. (2002). Training of working memory in children with ADHD. *Journal of Clinical and Experimental Neuropsychology.,* 24(6), 781-791.

Kofler, M. J., Rapport, M. D., Bolden, J., Sarver, D. E., Raiker, J. S., and Alderson, R. M. (2011). Working memory deficits and social problems in children with ADHD. *Journal of Abnormal Child Psychology,* 39(6), 805-817.

Kuriyan, A. B., Pelham, W. E., Jr., Molina, B. S., Waschbusch, D. A., Gnagy, E. M., Sibley, M. H., . . . Kent, K. M. (2013). Young adult educational and vocational outcomes of children diagnosed with ADHD. *Journal of Abnormal Child Psychology,* 41(1), 27-41.

Lan, X., Legare, C. H., Ponitz, C. C., Li, S., and Morrison, F. J. (2011). Investigating the links between the subcomponents of executive function and academic achievement: A cross-cultural analysis of Chinese and American preschoolers. *Journal of Experimental Child Psychology,* 108(3), 677-692.

Langberg, J. M., and Becker, S. P. (2012). Does long-term medication use improve the academic outcomes of youth with attention-deficit/hyperactivity disorder? *Clincal Child Family Psychology Review,* 15(3), 215-233.

Lara, C., Fayyad, J., de Graaf, R., Kessler, R. C., Aguilar-Gaxiola, S., Angermeyer, M., . . . Sampson, N. (2009). Childhood predictors of adult attention-deficit/hyperactivity disorder: Results from the World Health Organization World Mental Health Survey Initiative. *Biological Psychiatry,* 65(1), 46-54.

Mahone, E. M., Cirino, P. T., Cutting, L. E., Cerrone, P. M., Hagelthorn, K. M., Hiemenz, J. R., . . . Denckla, M. B. (2002). Validity of the Behavior Rating Inventory of executive function in children with ADHD and/or Tourette syndrome. *Archives of Clinical Neuropsychology,* 17, 643-662.

Mahone, E. M., and Hoffman, J. (2007). Behavior ratings of executive function among preschoolers with ADHD. *Clinical Neuropsychology,* 21(4), 569-586.

Martinussen, R., Hayden, J., Hogg-Johnson, S., and Tannock, R. (2005). A meta-analysis of working memory impairments in children with attention-deficit/hyperactivity disorder. *Journal of the American Academy of Child and Adolescent Psychiatry,* 44(4), 377-384.

Martinussen, R., and Tannock, R. (2006). Working memory impairments in children with attention-deficit hyperactivity disorder with and without comorbid language learning disorders. *Journal of Clinical and Experimental Neuropsychology,* 28(7), 1073-1094.

Massat, I., Slama, H., Kavec, M., Linotte, S., Mary, A., Baleriaux, D., . . . Peigneux, P. (2012). Working memory-related functional brain patterns in never medicated children with ADHD. *PLoS ONE,* 7(11), e49392.

Massetti, G. M., Lahey, B. B., Pelham, W. E., Loney, J., Ehrhardt, A., Lee, S. S., and Kipp, H. (2008). Academic achievement over 8 years among children who met modified criteria for attention-deficit/hyperactivity disorder at 4-6 years of age. *Journal of Abnormal Child Psychology,* 36(3), 399-410.

Maziade, M., Rouleau, N., Lee, B., Rogers, A., Davis, L., and Dickson, R. (2009). Atomoxetine and neuropsychological function in children with attention-deficit/hyperactivity disorder: Results of a pilot study. *Journal of Child and Adolescent Psychopharmacology,* 19(6), 709-718.

McCandless, S., and O' Laughlin, L. (2007). The Clinical utility of the Behavior Rating Inventory of Executive Function (BRIEF) in the diagnosis of ADHD. *Journal of Attention Disorders,* 10(4), 381-389.

McNab, F., Varrone, A., Farde, L., Jucaite, A., Bystritsky, P., Forssberg, H., and Klingberg, T. (2009). Changes in cortical dopamine D1 receptor binding associated with cognitive training. *Science,* 323(5915), 800-802.

Mehta, M. A., Goodyer, I. M., and Sahakian, B. J. (2004). Methylphenidate improves working memory and set-shifting in AD/HD: relationships to baseline memory capacity. *Journal of Child Psychology and Psychiatry,* 45(2), 293-305.

Mezzacappa, E., and Buckner, J. C. (2010). Working memory training for children with attention problems or hyperactivity: A school-based pilot study. *School Mental Health,* 2, 202-208.

Mills, K. L., Bathula, D., Dias, T. G., Iyer, S. P., Fenesy, M. C., Musser, E. D., . . . Fair, D. A. (2012). Altered cortico-striatal-thalamic connectivity in relation to spatial working memory capacity in children with ADHD. *Frontiers in Psychiatry,* 3, 2.

Miyake, A., and Shah, P. (Eds.). (1999). *Models of working memory: Mechanisms of active maintenance and executive control.* New York: Cambridge University press.

Murphy, K., and Barkley, R. A. (1996). Attention deficit hyperactivity disorder adults: comorbidities and adaptive impairments. *Comprehensive Psychiatry,* 37(6), 393-401.

O'Brien, J. W., Dowell, L. R., Mostofsky, S. H., Denckla, M. B., and Mahone, E. M. (2010). Neuropsychological profile of executive function in girls with attention-deficit/hyperactivity disorder. *Arch Clin Neuropsychol,* 25(7), 656-670.

Olesen, P. J., Westerberg, H., and Klingberg, T. (2004). Increased prefrontal and parietal activity after training of working memory. *Nature Neuroscience,* 7(1), 75-79.

Oord, S. V., Ponsioen, A. J., Geurts, H. M., Brink, E. L., and Prins, P. J. (in press). A pilot study of the efficacy of a computerized executive functioning remediation training with game elements for children with ADHD in an outpatient setting: Outcome on parent- and teacher-rated executive functioning and ADHD behavior. *Journal of Attention Disorders.*

Pietrzak, R. H., Mollica, C. M., Maruff, P., and Snyder, P. J. (2006). Cognitive effects of immediate-release methylphenidate in children with attention-deficit/hyperactivity disorder. *Neuroscience and Biobehavioral Reviews,* 30(8), 1225-1245.

Prasad, V., Brogan, E., Mulvaney, C., Grainge, M., Stanton, W., and Sayal, K. (2013). How effective are drug treatments for children with ADHD at improving on-task behaviour and academic achievement in the school classroom? A systematic review and meta-analysis. *European Child and Adolescent Psychiatry*, 22(4), 203-216.

Prins, P. J., Dovis, S., Ponsioen, A., ten Brink, E., and van der Oord, S. (2011). Does computerized working memory training with game elements enhance motivation and training efficacy in children with ADHD? *Cyberpsychology, Behavior, and Social Networking*, 14(3), 115-122.

Raiker, J. S., Rapport, M. D., Kofler, M. J., and Sarver, D. E. (2012). Objectively-measured impulsivity and attention-deficit/hyperactivity disorder (ADHD): Testing competing predictions from the working memory and behavioral inhibition models of ADHD. *Journal of Abnormal Child Psychology*, 40(5), 699-713.

Rapport, M. D., Alderson, R. M., Kofler, M. J., Sarver, D. E., Bolden, J., and Sims, V. (2008). Working memory deficits in boys with attention-deficit/hyperactivity disorder (ADHD): The contribution of central executive and subsystem processes. *Journal of Abnormal Child Psychology*, 36(6), 825-837.

Rapport, M. D., Chung, K. M., Shore, G., and Isaacs, P. (2001). A conceptual model of child psychopathology: implications for understanding attention deficit hyperactivity disorder and treatment efficacy. *Journal of Clinical Child Psychology*, 30(1), 48-58.

Rhodes, S. M., Park, J., Seth, S., and Coghill, D. R. (2012). A comprehensive investigation of memory impairment in attention deficit hyperactivity disorder and oppositional defiant disorder. *Journal of Child Psychology and Psychiatry*, 53(2), 128-137.

Richards, T., Deffenbacher, J., and Rosen, L. (2002). Driving anger and other driving-related behaviors in high and low ADHD symptom college students. *Journal of Attention Disorders*, 6(1), 25-38.

Rogers, M., Hwang, H., Toplak, M., Weiss, M., and Tannock, R. (2011). Inattention, working memory, and academic achievement in adolescents referred for attention deficit/hyperactivity disorder (ADHD). *Child Neuropsychology*, 17(5), 444-458.

Rommelse, N. N., Van der Stigchel, S., Witlox, J., Geldof, C., Deijen, J. B., Theeuwes, J., . . . Sergeant, J. A. (2008). Deficits in visuo-spatial working memory, inhibition and oculomotor control in boys with ADHD and their non-affected brothers. *Journal of Neural Transmission*, 115(2), 249-260.

Rotenberg-Shpigelman, S., Rapaport, R., Stern, A., and Hartmen-Maeir, A. (2008). Content validity and internal consistency reliability of the Behavior Rating Inventory of Executive Function - Adult Version (BRIEF-A) in Israeli adults with attention-deficit/hyperactivity disorder. *Israeli Journal of Occupational Therapy*, 17(2), 77-96.

Roth, R. M., Isquith, P. K., and Gioia, G. A. (2005). Behavior Rating Inventory of Executive Function - Adult Version (BRIEF-A). Lutz, FL: Psychological Assessment Resources.

Rubio, B., Hernandez, S., Verche, E., Martin, R., and Gonzalez-Perez, P. (2011). A pilot study: differential effects of methylphenidate-OROS on working memory and attention functions in children with attention-deficit/hyperactivity disorder with and without behavioural comorbidities. *Attention Deficit Hyperactivity Disorder*, 3(1), 13-20.

Shah, P., Buschkuehl, M., and Jonides, J. (2012). Cognitive training in ADHD: The importance of individual differences. *Journal of Applied Research in Memory and Cognition*, 1, 204-205.

Shaw, M., Hodgkins, P., Caci, H., Young, S., Kahle, J., Woods, A. G., and Arnold, L. E. (2012). A systematic review and analysis of long-term outcomes in attention deficit hyperactivity disorder: effects of treatment and non-treatment. *BMC Medicine,* 10(1), 99.

Shiels, K., Hawk, L. W., Jr., Lysczek, C. L., Tannock, R., Pelham, W. E., Jr., Spencer, S. V., Gangloff, B. P., and Waschbusch, D. A. (2008). The effects of incentives on visual-spatial working memory in children with attention-deficit/hyperactivity disorder. *Journal of Abnormal Child Psychology,* 36(6), 903-913.

Soderqvist, S., Nutley, S. B., Peyrard-Janvid, M., Matsson, H., Humphreys, K., Kere, J., and Klingberg, T. (2012). Dopamine, working memory, and training induced plasticity: Implications for developmental research. *Developmental Psychology,* 48(3), 836-843.

Sohlberg, M. M., and Mateer, C. A. (2001). Cognitive rehabilitation: An integrative neuropsychological approach (2nd ed.). New York: The Guilford Press.

Sowerby, P., Seal, S., and Tripp, G. (2011). Working memory deficits in ADHD: The contribution of age, learning/language difficulties, and task parameters. *Journal of Attention Disorders,* 15(6), 461-472.

Strand, M. T., Hawk, L. W., Jr., Bubnik, M., Shiels, K., Pelham, W. E., Jr., and Waxmonsky, J. G. (2012). Improving working memory in children with attention-deficit/hyperactivity disorder: The separate and combined effects of incentives and stimulant medication. *Journal of Abnormal Child Psychology,* 40(7), 1193-1207.

Stuss, D. T., Winocur, G., and Robertson, I. H. (Eds.). (2010). Cognitive neurorehabilitation: Evidence and application (2nd ed.). New York Cambridge University Press.

Sumner, C. R., Gathercole, S., Greenbaum, M., Rubin, R., Williams, D., Hollandbeck, M., and Wietecha, L. (2009). Atomoxetine for the treatment of attention-deficit/hyperactivity disorder (ADHD) in children with ADHD and dyslexia. *Child Adolescent Psychiatry Mental Health,* 3, 40.

Surman, C. B., Hammerness, P. G., Petty, C., Spencer, T., Doyle, R., Napolean, S., . . . Biederman, J. (in press). A pilot open label prospective study of memantine monotherapy in adults with ADHD. *World Journal of Biological Psychiatry.*

Swanson, J., Baler, R. D., and Volkow, N. D. (2011). Understanding the effects of stimulant medications on cognition in individuals with attention-deficit hyperactivity disorder: A decade of progress. *Neuropsychopharmacology,* 36(1), 207-226.

Thompson, A. L., Molina, B. S. G., Pelham, W., Jr., and Gnagy, E. M. (2007). Risky driving in adolescents and young adults with childhood ADHD. *Journal of Pediatric Psychology,* 32(7), 745-759.

Toplak, M. E., Bucciarelli, S. M., Jain, U., and Tannock, R. (2008). Executive functions: performance-based measures and the Behavior Rating Inventory of Executive Function (BRIEF) in adolescents with attention deficit/hyperactivity disorder (ADHD). *Child Neuropsychology,* 15(1), 53-72.

Turner, D. C., Blackwell, A. D., Dowson, J. H., McLean, A., and Sahakian, B. J. (2005). Neurocognitive effects of methylphenidate in adult attention-deficit/hyperactivity disorder. *Psychopharmacology,* 178(2-3), 286-295.

Valera, E. M., Faraone, S. V., Biederman, J., Poldrack, R. A., and Seidman, L. J. (2005). Functional neuroanatomy of working memory in adults with attention-deficit/hyperactivity disorder. *Biological Psychiatry,* 57(5), 439-447.

Wager, T. D., and Smith, E. E. (2003). Neuroimaging studies of working memory: A meta-analysis. *Cognitive, Affective and Behavioral Neuroscience,* 3(4), 255-274.

Westerberg, H., and Klingberg, T. (2007). Changes in cortical activity after training of working memory: A single-subject analysis. *Physiology and Behavior,* 92(1-2), 186-192.

Willcutt, E. G., Doyle, A. E., Nigg, J. T., Faraone, S. V., and Pennington, B. F. (2005). Validity of the executive function theory of attention-deficit/hyperactivity disorder: A meta-analytic review. *Biological Psychiatry,* 57(11), 1336-1346.

Wilson, B. A. (Ed.). (2003). Neuropsychological rehabilitation: Theory and practice. New York: Taylor and Francis.

Wolf, R. C., Plichta, M. M., Sambataro, F., Fallgatter, A. J., Jacob, C., Lesch, K. P., . . . Vasic, N. (2009). Regional brain activation changes and abnormal functional connectivity of the ventrolateral prefrontal cortex during working memory processing in adults with attention-deficit/hyperactivity disorder. *Human Brain Mapping,* 30(7), 2252-2266.

Wong, C. G., and Stevens, M. C. (2012). The effects of stimulant medication on working memory functional connectivity in attention-deficit/hyperactivity disorder. *Biological Psychiatry,* 71(5), 458-466.

Yang, L., Cao, Q., Shuai, L., Li, H., Chan, R. C., and Wang, Y. (2011). Comparative study of OROS-MPH and atomoxetine on executive function improvement in ADHD: A randomized controlled trial. *International Journal of Neuropsychopharmacology,* 15(1), 15-26.

Young, C. E., and Yang, C. R. (2005). Dopamine D1-like receptor modulates layer- and frequency-specific short-term synaptic plasticity in rat prefrontal cortical neurons. *European Journal of Neuroscience,* 21(12), 3310-3320.

In: Working Memory
Editor: Helen St. Clair-Thompson

ISBN: 978-1-62618-927-0
© 2013 Nova Science Publishers, Inc.

Chapter 13

A DUAL-COMPONENT ANALYSIS OF WORKING MEMORY TRAINING

Bradley S. Gibson and Dawn M. Gondoli
University of Notre Dame, South Bend, IN, US

ABSTRACT

Working memory (WM) capacity constrains reasoning and learning abilities and students who lack adequate WM capacity do not meet benchmark standards in core academic subjects such as reading and math. This association between WM capacity and academic competence has given rise to the expectation that WM capacity is a malleable factor. Indeed, several adaptive training regimens have attempted to improve student outcomes in reasoning and learning by increasing WM capacity. Unfortunately, there is growing consensus that these particular training regimens neither increase WM capacity adequately nor improve higher-level cognitive or academic abilities. But, before WM capacity can be dismissed as a malleable factor, it must first be shown that these particular training regimens have targeted the proper theoretical mechanisms. For instance, according to the dual-component theory, there are two critical components of WM capacity—an attention component and a memory component—that account for the relation between WM capacity and higher-level cognitive abilities. However, we have recently shown that existing regimens only target one of the two components (the attention component), which may explain why they have proven to be ineffective. There is, therefore, a critical need to create and test new training regimens that are capable of targeting both components of WM capacity. This chapter reviews recent evidence from our lab that has led to the development of a novel training regimen that can target and enhance both the attention and memory components of WM capacity. We argue that this development is significant because it will allow researchers to test the central hypothesis that improvements in student reasoning and learning will be moderated by the number of WM components that are targeted and enhanced by the training regimen.

INTRODUCTION

WM is an executive function that constrains many higher-level cognitive and academic abilities. For instance, individual differences in WM capacity have been found to reliably predict a range of higher-level cognitive and academic abilities in young adults, including reading comprehension abilities (Daneman and Carpenter, 1980; Turner and Engle, 1989), fluid IQ (Engle, Tuholski, Laughlin, and Conway, 1999; Kane and Engle, 2002; Kane et al., 2004; Unsworth and Spillers, 2010), and performance on college entrance exams (Cowan et al., 2005; Turner and Engle, 1989). Likewise, individual differences in WM capacity have also been found to reliably predict a range of higher-level cognitive and academic abilities in children, including fluid IQ (Engel de Abreu, Conway, and Gathercole, 2010), verbal and mathematical aptitude (Cowan et al, 2005; Gathercole and Pickering, 2000; Swanson and Beebe-Frankenberger, 2004), as well as the rate at which children develop syntactic knowledge and reading ability (Engel de Abreu, Gathercole, and Martin, 2011).

Altogether, individual difference studies have suggested that children and adults with higher WM capacities typically demonstrate greater reasoning and learning abilities than children and adults with lower WM capacities (Bull and Scherif, 2001; Cain and Oakhill, 2006; Engle, Carullo, and Collins, 1991; Gathercole, Brown, and Pickering, 2003; Geary, Hoard, Nugent, and Byrd-Craven, 2007; Swanson, Zheng, and Jerman, 2009). In fact, there has been growing interest in the subset of students who are in the lowest 10-15 percentile of WM capacity because these students appear to manifest a common pattern of maladaptive classroom behaviors that put them at risk for academic problems. For instance, students with low WM tend to be inattentive, have poor study skills, are poor note-takers, and are unorganized. Most importantly, students with low WM have poor academic outcomes in core curriculum subjects such as reading and math (Alloway, Elliott, and Place, 2010; Alloway, Gathercole, Kirkwood, and Elliott, 2009; Alloway, Gathercole, and Elliott, 2010; DiPerna and Elliott, 2000; DuPaul et al., 2004).

Many students with low WM capacity also meet criteria for various exceptionalities (e.g., autism spectrum disorder, communication disorder, or learning disability) or other medically defined disorders that may not be formally delineated within the educational exceptionalities (e.g., attention-deficit/hyperactivity disorder, extreme prematurity, traumatic brain injury). However, not all students with these specific disabilities have low WM (Archibald and Joanisse, 2009; Dahlin, 2010; Kyttälä, Aunio, and Hautamäki, 2010; Loo et al., 2007; Montgomery, Magimairaj, and Finney, 2010; Willcutt, Doyle, Nigg, Faraone, and Pennington, 2005), suggesting that the identification of students on the basis of low WM may be important in its own right (see also, Dahlin, 2010; Holmes, Gathercole, and Dunning, 2009; Holmes et al., 2010; Klingberg et al., 2005; Mezzacappa and Buckner, 2010; St Clair-Thompson, Stevens, Hunt, and Bolder, 2010; Thorell, Lindqvist, Nutley, Bohlin, and Klingberg, 2009).

IS WM CAPACITY A MALLEABLE FACTOR?

Because the capacity to reason and learn appears to be constrained by the capacity of WM, there has been great interest in the question of whether WM capacity is malleable and capable of being increased by adaptive training regimens. Indeed, a debate has erupted over the past decade about whether adaptive WM training regimens can enhance WM capacity and associated higher-order cognitive and adaptive functioning. On the one hand, several empirical studies have been interpreted to suggest that training-induced increases in WM capacity can occur and can also be accompanied by improvements in fluid intelligence (Jaeggi, Buschkuehl, Jonides, and Perrig, 2008; Jaeggi, Buschkuehl, Jonides, and Shah, 2011; Klingberg et al., 2005), reading comprehension (Chein and Morrison, 2010; Dahlin, 2010), math competence (Holmes et al., 2009), and ADHD symptoms (Gibson, Gondoli, Johnson, Steeger, and Morrissey, 2011; Klingberg et al., 2005). These positive findings have been emphasized in several recent reviews that have been optimistic about the possibility that training regimens can enhance WM capacity and associated higher-level abilities (Buschkuehl and Jaeggi, 2010; Diamond and Lee, 2011; Klingberg, 2010; Morrison and Chein, 2011).

On the other hand, several other, more comprehensive, reviews have been considerably more pessimistic about the effectiveness of WM training (Melby-Lervag and Hulme, 2012; Shipstead, Hicks, and Engel, 2012; Shipstead, Redick, and Engle, 2010; Shipstead, Redick, and Engle, 2012). For instance, on the basis of their comprehensive meta-analysis of WM training in children and adults, Melby-Lervag and Hulme (2012) concluded that although "[c]urrent training programs yield reliable, short-term improvements on both verbal and nonverbal working memory tasks…there is no evidence that working memory training produces generalized gains to the other skills that have been investigated (verbal ability, word decoding, or arithmetic), even when assessments take place immediately after training" (p. 12). Likewise, on the basis of their comprehensive narrative review of WM training in children and adults, Shipstead et al., (2012) also concluded that there is no evidence that WM training produces generalized gains to higher-level cognitive and academic abilities (see also, Shipstead et al., 2010; Shipstead et al., 2012). Furthermore, Shipstead et al. (2010) also questioned whether WM training can produce short-term improvements of verbal and nonverbal WM, based on a variety of concerns about the measurement of these outcomes, the theoretical conception of WM, and the use of appropriate control groups (see also, Shipstead et al., 2012).

Based on the conclusions of these more comprehensive reviews (Melby-Lervag and Hulme, 2012; Shipstead et al., 2010; Shipstead et al., 2012), there appears to be little evidence that WM capacity is malleable or that WM training can influence student reasoning and learning. However, before these conclusions can be accepted with confidence, it is first necessary to explore an alternative explanation: namely, that existing training regimens have not been optimally designed.

NEW THEORETICALLY-INSPIRED WM TRAINING REGIMES MUST BE CREATED AND EXPLORED

We have recently argued that the full potential of WM training has not yet been adequately tested because existing regimens do not target the proper theoretical mechanisms (Gibson, Gondoli et al., 2012). Likewise, Melby-Lervag and Hulme (2012) state that existing regimens "do not appear to rest on any detailed task analysis or theoretical account of the mechanisms by which such adaptive training regimes would be expected to improve working memory capacity. Rather, these programs seem to be based on what might be seen as a fairly naïve 'physical-energetic' model such that repeatedly 'loading' a limited cognitive resource will lead to it increasing in capacity, perhaps somewhat analogously to strengthening a muscle by repeated use" (p. 3). Similarly, Shipstead et al. (2012) state, "we do not rule out the possibility that WM training could be made effective. The largest issue seems to be that, while there is logic to WM training (increase WM and improve related abilities), this literature is still struggling to find a theory. Specifically, it is important that research move beyond the desire to show that broad change can be realized through a month of training on a limited set of tasks" (p. 22).

According to Gibson, Gondoli et al. (2012), theories of WM capacity have important implications for the design of WM training regimens because it is in virtue of these theories that we come to understand (1) which components of WM distinguish high-capacity individuals from low-capacity individuals, (2) which components are important for supporting higher-level cognitive and academic abilities, and (3) how these components should be measured. In short, if the primary goal of WM training is to increase an individual's WM capacity from low to high so that reasoning and learning abilities can be improved, then theories of WM capacity can indicate which components of WM capacity are the most important to target, and they can also indicate how those components should be trained.

WHICH COMPONENTS OF WM CAPACITY DISTINGUISH HIGH- AND LOW-CAPACITY INDIVIDUALS?

The main purpose of WM training is to increase the capacity of WM. But, what is the nature of this capacity? Answers to this question are provided by theories that attempt to identify the critical mechanisms or components of WM capacity. In this chapter, we distinguish between one-component and two-component theories of WM capacity. The most prominent one-component theories have typically included flexible attention mechanisms such as the "focus of attention" (Cowan, 1995; 2001; Cowan et al., 2005) or "executive attention" (Kane and Engle, 2002) that allow individuals to actively maintain a limited amount of goal-relevant information in "primary memory" (PM), especially in the presence of distraction. Although both attention mechanisms were postulated to explain how information is actively maintained in WM, the focus of attention emphasized the maximum number of items that can be maintained at one time, whereas executive attention emphasized the ability to maintain goal-relevant information in the face of irrelevant distraction.

The most prominent two-component theories have typically included these flexible attention mechanisms, but they have also included memory mechanisms such as "cue-dependent retrieval" that allow individuals to retrieve goal-relevant information from "secondary memory" (SM), after that information has been lost from PM (Mogle, Lovett, Stawski, and Sliwinski, 2008; Unsworth and Engle, 2007a; 2007b; Unsworth and Spillers, 2010). Current theories construe recall from SM as a multi-step process. For instance, Unsworth (2007; 2009) has construed recall from SM in terms of three parameters: the size of the search set, the recovery of potential targets from this set, and error monitoring. In this chapter, we refer to the attention mechanisms as the PM component of WM capacity, and we will refer to the memory mechanisms as the SM component of WM capacity.

Empirical studies that have investigated whether one or two components are necessary to distinguish high- and low-capacity individuals have typically used an extreme-groups approach. In these studies, individuals who fall in the top and bottom quartiles of overall WM capacity are compared on measures that are specifically designed to isolate the PM and SM components of WM capacity (Unsworth and Engle, 2007a). These findings have shown that high-capacity individuals differ from low-capacity individuals in terms of both PM and SM abilities. Based on this evidence, Unsworth and Engle concluded that low-capacity individuals were less likely to maintain goal-relevant information in PM than high-capacity individuals, and they were also less proficient at using cues to retrieve goal-relevant information from SM, after that information was lost from PM (see also, Gibson, Gondoli, Flies, Dobrzenski, and Unsworth, 2010; Unsworth, 2007). Thus, both PM and SM components of WM capacity appear to be necessary to distinguish low-capacity individuals from high-capacity individuals. This notion has come to be called the "dual-component theory" of WM capacity.

WHICH COMPONENTS OF WM CAPACITY ARE IMPORTANT FOR SUPPORTING HIGHER-LEVEL COGNITIVE AND ACADEMIC ABILITIES?

Further evidence for the dual-component theory has come from empirical studies that investigated whether both components are related to higher-level cognitive and academic abilities. These studies have typically used a latent variable approach in which the full range of individual PM and SM abilities is investigated within the context of structural equation modeling (Mogle, Lovett, Stawski, and Sliwinski, 2008; Unsworth and Engle, 2007b; Unsworth and Spillers, 2010). For instance, Unsworth and Engle (2007b) showed that both the PM and SM factors accounted for significant variance in SAT and fluid IQ; however, the SM factor accounted uniquely for more variance than the PM factor. Similarly, Mogle et al. (2008) also showed that SM ability was more important than PM capacity for explaining the relation between WM capacity and fluid IQ. However, Unsworth and his colleagues (Unsworth, Brewer, and Spillers, 2009; Unsworth and Spillers, 2010) have recently provided evidence that both PM and SM ability are equally important for explaining the relation between WM capacity and fluid IQ.

In addition, using tasks that focused solely on the SM component of WM capacity, Unsworth (2009) operationalized the three parameters of the SM component—i.e., the size of the search set, the recovery of potential targets from this set, and error monitoring—in terms of recall latency, recall accuracy, and intrusion errors, respectively, in order to examine how pre-existing individual differences in these three parameters related to pre-existing differences in WM capacity and fluid IQ. The main findings suggested that pre-existing differences in WM capacity and fluid IQ were primarily related to differences in recall latency and recall accuracy, but not intrusion errors. In particular, individuals with higher WM capacity and fluid IQ tended to have faster recall latencies (indicating the use of smaller search sets likely containing less irrelevant information) and higher recall accuracies (indicating better recovery of potential targets from the search set).

In short, these findings suggest that both the PM and SM components are important for explaining individual differences in WM capacity, reasoning, and learning. Hence, these findings have important implications for the design of WM training regimens because they suggest that a regimen that can target both the PM and SM components of WM capacity should have stronger effects on higher-level cognitive and academic abilities than a regimen that can target only one or neither of the two components.

How Should the PM and SM Components of WM Capacity be Measured? Implications for Training

Consideration of the measurement of WM capacity is important because the tasks that have been developed to measure WM capacity are also used to train WM capacity. The measurement of WM capacity has generally relied on the span task. Span tasks typically come in two varieties: simple span and complex span. Simple span tasks typically involve the presentation of lists of various lengths that participants attempt to recall in forward serial order immediately following the conclusion of the list (e.g., forward word span or forward digit span). In contrast, complex span tasks typically involve the performance of two tasks on each trial. For instance, in the operation span task (Turner and Engle, 1989), participants must first solve a mathematical operation and then attempt to store a simultaneously (or sequentially) presented list item on each trial. As in simple span tasks, lists of various lengths are presented that participants attempt to recall in forward serial order immediately following the conclusion of the list.

There has been a debate about whether the simple span task or complex span task is a better measure of WM capacity. Initial support for the importance of complex span tasks over simple span tasks came from studies which showed that individual differences in complex span performance correlated more highly with higher-level cognitive abilities such as reading comprehension, SAT, and fluid intelligence than did individual differences in simple span performance (see e.g., Daneman and Carpenter, 1980; Engle et al., 1999; Turner and Engle, 1989). These findings were originally interpreted within a one-component (attention-based) theory of WM capacity. For instance, according to Cowan et al. (2005), the critical feature of complex span tasks is that the processing component suppresses rehearsal strategies that likely operate in simple span tasks. As a result, complex span tasks may provide a purer measure of the focus of attention (the number of items that can be held active at any given point in time). Likewise, according to Kane and his colleagues (Kane and Engle, 2002; Kane et al., 2004), the critical feature of complex span tasks is that the processing component represents a source of interference that is not present in simple span tasks. As a result, complex span tasks may provide a purer measure of domain-general executive attention processes that enable individuals to maintain goal-relevant information in the presence of interference whereas simple span tasks generally do not.

More recently, Unsworth and Engle (2007b) used a two-component theory of WM capacity to explain the stronger correlation that is typically found between complex span tasks and higher-order cognition than between simple span tasks and higher-order cognition. In particular, they argued that simple span and complex span tasks both measure the same processes (rehearsal, maintenance, and updating in PM, and retrieval from SM), albeit to different extents. For instance, using confirmatory factor analysis, Unsworth and Engle showed that simple span tasks such as forward word span loaded more highly on the PM factor than on the SM factor; whereas, complex span tasks such as operation span loaded more highly on the SM factor than on the PM factor.

According to Unsworth and Engle (2007b), complex span tasks that require dual-task performance (e.g., operation span) provide better measures of SM ability than PM capacity because the processing task causes all but the last of the to-be-remembered list items to be displaced from PM into SM. As a result, successful recall in complex span tasks mostly reflects the retrieval of information from SM. In contrast, simple span tasks provide better measures of PM capacity than SM ability because the displacement of items from PM into SM only occurs with relatively long list lengths in these tasks (i.e., with list lengths that exceed the storage capacity of PM). As a result, successful recall in simple span tasks mostly reflects the unloading of information that is actively maintained in PM, at least when list-length is relatively short. However, successful recall in simple span tasks may increasingly measure SM ability (as opposed to PM capacity) as list-length increases beyond the capacity of PM. Indeed, Unsworth and Engle (2006) showed that simple span tasks can predict fluid IQ scores just as well as complex span tasks when list-length exceeds the capacity of PM (greater than 4 items) because the simple span task now measures both the PM and SM components of WM capacity. Thus, according to this analysis, both simple span and complex span tasks appear capable of targeting the PM and SM components of WM capacity under certain conditions. Below we will consider the effects of using simple vs. complex span tasks as exercises in WM training regimens. In addition, we will also consider the effects of using different adaptive training algorithms that adjust the length of the memory lists used during WM training on an individual-by-individual basis.

Furthermore, it is also important to point out that Unsworth and Engle (2007a) have used immediate free recall (IFR) tasks to assess the individual PM and SM components of WM capacity. These tasks are similar to simple span tasks, except that relatively long lists are presented on every trial (e.g., 12 items) and the items can be recalled in any order (see the Research Plan for further details). According to Unsworth and Engle, IFR tasks are valid measures of WM capacity. In their re-analysis of Engle et al.'s (1999) structural equation model of the relation between WM capacity, fluid intelligence, and scholastic aptitude, Unsworth and Engle showed that performance on a verbal IFR task loaded just as highly (0.77) on the latent construct of WM capacity as performance on three more traditional complex span tasks did: operation span (0.77), reading span (0.58), and counting span (0.62).

More importantly, IFR tasks may be better suited for assessing recall from PM and SM than complex or simple span tasks because IFR tasks can provide separate measures of each component, whereas complex and simple span tasks typically provide a single measure that may reflect contributions from both components. In particular, performance on IFR tasks can be divided into recency and pre-recency portions: Typically, individuals are better at recalling the last few presented (recency) items than they are at recalling the earlier presented (pre-recency) items because the recency items can be actively maintained and then simply unloaded from PM whereas the pre-recency items need to be encoded and then retrieved by means of a probabilistic search through SM.

APPLICATION OF THE DUAL-COMPONENT THEORY OF WM CAPACITY TO EXISTING WM TRAINING REGIMENS

Recently, we have used the dual-component theory of WM capacity to provide a detailed task analysis and theoretical account of the mechanisms by which one existing adaptive training intervention influences WM capacity (Gibson et al., 2011; Gibson, Gondoli et al., 2012; Gibson, Kronenberger et al., 2012). In our initial study, we investigated whether the PM component, the SM component, or both components of WM capacity could be enhanced by a span-based intervention known as "Cogmed-RM" (Gibson et al., 2011). We focused on this intervention because it is generally considered to be the best-known WM training intervention in existence today, having been widely-used in both schools and clinics around the world, as well as being either a major focus (Melby-Lervag and Hulme, 2012; Shipstead et al., 2012), or the sole focus (Shipstead et al., 2012) of previous reviews. In addition, Cogmed-RM utilizes both verbal and spatial span tasks as training exercises. Because the dual-component theory is largely rooted in the analysis of span task performance, this theory can be more directly applied to this intervention than it could to other training regimens that utilize other training exercises such as the n-back task that are not clearly related to span task performance (Jaeggi et al., 2008).

According to Holmes et al. (2009), of the 10 training exercises contained within Cogmed-RM, three involved the simple storage of verbal information, two involved the simple storage of spatial information, two involved the active updating and manipulation of verbal information, and three involved the active updating and manipulation of spatial information. According to the analysis provided by Unsworth and Engle (2007b), these training tasks may engage the PM component more than the SM component of WM capacity, though these tasks may have the potential to engage the SM component if training routinely involves list lengths that exceed the capacity of PM. In addition, other research has suggested that spatial simple span tasks may function more like complex span tasks than verbal simple span tasks (Kane et al., 2004; Miyake, Friedman, Rettinger, Shah, and Hegarty, 2001; Oberauer, 2005; Shah and Miyake, 1996). Thus, it is possible that adaptive training with spatial simple span exercises may enhance the SM component more than adaptive training with verbal simple span exercises. Accordingly, the Cogmed exercises were divided into two separate training conditions in Gibson et al.'s (2011) study—a verbal training condition ($N = 20$) and a spatial training condition ($N = 17$)—to examine whether spatial training might engage the SM component more than verbal training using a randomized, controlled design. The sample consisted of adolescents (11-14 years of age) with ADHD.

Following Unsworth and Engle (2007a), serial position effects and estimates of PM and SM capacity were derived from performance on verbal and spatial IFR tasks that were administered before and immediately following at least 20 days of training. The main findings showed that training selectively improved the PM component of WM capacity ($d = 0.52$), but not the SM component of WM capacity ($d = 0.15$), regardless of training condition. The finding that span-based training only targets a single component of WM capacity is important because it suggests that this training is not as potent as it could be (Unsworth and Spillers, 2010), and may explain why previous training studies that have used span-based training tasks have resulted in only weak or ineffective training of WM capacity and associated higher-level abilities (Shipstead, Hicks et al., 2012). Hence, according to the dual-component theory of WM capacity, the full potential of span-based training regimens has not yet been adequately tested.

Accordingly, the main purpose of our recent WM training research has been to translate the dual-component theory of WM into a novel WM training regimen that can target both the PM and SM components of WM capacity. In agreement with this goal, Shipstead et al. (2012) concluded, "[o]n the basis of the empirical work of Unsworth and Engle …which emphasizes the importance of retrieval from secondary memory, Gibson et al. … have begun modifying Cogmed [i.e., span-based training regimens] in an attempt to produce specific desired training effects (by creating greater need for retrieval). Under mounting evidence that Cogmed and other training programs have not lived up to the promise of early studies … these endeavors represent a sensible course of action" (p. 22). Translating the dual-component theory of WM capacity into a more potent training regimen will require specific hypotheses about how the SM component of WM capacity is related to aspects of adaptive training that are under the control of the experimenter and/or educational system. Two such hypotheses are examined in the next two sections.

TARGETING THE SM COMPONENT OF WM CAPACITY: EXAMINATION OF EXERCISE TYPE

Gibson et al. (2011) found that span-based training regimens enhanced the maintenance of information in PM, but not the retrieval of information from SM, and they hypothesized that this pattern of results may have occurred because the exercises they used were primarily simple span exercises as opposed to complex span exercises. Accordingly, Gibson, Kronenberger et al. (2012) investigated whether training with complex span exercises would target the SM component more than training with simple span exercises. Chein and Morrison (2010) also developed training exercises based on complex span tasks, but they did not measure the PM and SM components of WM capacity.

The complex span exercises were created by inserting additional processing tasks between to-be-remembered list items in a critical subset of both verbal and spatial exercises, similar to the operation span (Turner and Engle, 1989) and symmetry span tasks (Kane et al., 2004), respectively, and two separate training conditions were compared: a simple span training condition ($N = 31$) and a complex span training condition ($N = 30$) using a randomized, controlled design. The sample consisted of adolescents (9 to 16 years of age) without regard for diagnostic status.

If inserting an additional processing task increases the probability that list items are lost from PM (Unsworth and Engle, 2007b), then training with adaptive complex span exercises should target the SM component of WM capacity more than training with adaptive simple span exercises. Thus, the SM component of WM capacity should be enhanced to a greater extent following training in the complex span training condition than in the simple span training condition. For this reason, the complex span training condition was construed as the treatment condition in this study whereas the simple span training condition was construed as the control condition. Construing the simple span training condition as a control condition may seem unusual given that it is an active training condition. However, the simple span training condition is arguably the most optimal control condition for present purposes because the active component of this condition should induce the same motivation and expectations as the complex span training condition while also having no effect on the SM component.

Although the simple span training condition could be construed as a control condition with respect to SM capacity, it could not be construed as a control condition with respect to PM capacity because the PM component of WM capacity was expected to be enhanced more or less equally across the two training conditions. Nevertheless, no further attempt was made to clarify the nature of the changes observed in the PM component of WM capacity because the main purpose of this study was to determine if training with complex span exercises could target the SM component of WM capacity.

Following Gibson et al. (2011), serial position effects and estimates of PM and SM capacity were derived from performance on verbal and spatial IFR tasks. As expected, the findings reported by Gibson, Kronenberger et al. (2012) showed that the simple span training condition selectively improved the PM component of WM capacity ($d = 0.36$), but not the SM component of WM capacity ($d = 0.04$). These findings replicated the findings reported by Gibson et al. (2011). However, despite evidence that the complex span exercises were more distracting than the simple span exercises during training, the same pattern of results was also observed in the complex span training condition: Namely, the PM component of WM capacity was improved ($d = 0.47$), but the SM component of WM capacity was not ($d = 0.03$). Based on these findings, Gibson, Kronenberger et al. (2012) concluded that training with complex span exercises is not sufficient to target the SM component of WM capacity.

TARGETING THE SM COMPONENT OF WM CAPACITY: EXAMINATION OF RECALL ACCURACY THRESHOLD

Although the use of complex span tasks may increase the probability that any given item is lost from PM during training, satisfaction of this criterion alone does not guarantee that trainees are given adequate opportunities to practice retrieving this information from SM. Rather, providing adequate opportunities to practice retrieving information from SM may require further consideration of how the span length of the adaptive exercises is adjusted on a trial-by-trial basis to match the capacity of the trainee.

According to the dual-component theory, overall WM capacity arises from two sources: the capacity of PM and the efficiency with which information is encoded and retrieved from SM. And, there are at least two reasons to suspect that the adaptive algorithm used in previous training regimens is biased to target PM capacity, but not SM ability. First, the recall accuracy threshold used to adjust list length in previous span-based training regimens (Chein and Morrison, 2010; Gibson et al., 2011; Holmes et al., 2009; Klingberg et al., 2005) has been universally set at 100%. As a result, the length of the upcoming list will not increase until the trainee can consistently recall all the items on the current list with perfect accuracy. Second, recall from SM tends to be less accurate than recall from PM (Unsworth and Engle, 2007a). This is because recall from PM has been construed as simply "unloading" the contents of PM whereas recall from SM involves a probabilistic search through a set of both relevant and irrelevant items (Unsworth, 2007; Unsworth et al., 2009; Unsworth and Engle, 2007a).

If recall from SM is less accurate than recall from PM, then the recall accuracy threshold may constrain the engagement of the SM component. For instance, consider an individual who is training with a 100% recall accuracy threshold, and consider that this individual has just encountered a list that exceeded the capacity of PM by one item. Let us suppose further that this individual was able to recall all the items that were being maintained in PM with perfect accuracy, but failed to recall the one item that was lost from PM and had to be retrieved from SM. Because list length is contingent on perfect recall in this context, the length of the next list will be decreased by one item. In this way, a 100% recall accuracy threshold may enable this individual to train at the maximal (or near maximal) capacity of PM, without providing much opportunity to train the SM component. In contrast, now suppose that this individual had been training with a lower recall accuracy threshold. Although, they failed to correctly recall the one item that was lost from PM, the length of the next list will not decrease, but rather will continue to increase until this individual is unable to satisfy the lower recall accuracy threshold. Consequently, this individual will now be given more opportunity to practice retrieving list items from SM, and as a result, his/her ability to retrieve may improve and SM ability may increase. Thus, increased engagement of the SM component during training may require decreasing the recall accuracy threshold from 100% to a lower value. A decrease in the recall accuracy threshold will likely elicit more retrieval from SM before recall is terminated on any given training trial, and it will also ensure that list length is determined more by the limitations of SM ability than by the limitations of PM capacity.

Gibson, Gondoli, Kronenberger, Johnson, Steeger, and Morrissey (2013) have recently investigated this hypothesis. In this study, the recall accuracy threshold decreased as list length increased, averaging approximately 80%. If lowering the recall accuracy threshold is sufficient to target the SM component, then significant enhancement of SM ability should be observed across time. In addition, this lower recall accuracy threshold was implemented within both the simple span ($N = 10$) and complex span ($N = 9$) training conditions used by Gibson, Kronenberger et al. (2012) in order to examine whether it might interact with exercise type. If lowering the recall accuracy threshold interacts with exercise type, then greater enhancement of SM abilities should be observed across time in the complex span training condition than in the simple span training condition. Furthermore, significant enhancement of PM capacity should also be observed across time in the present study regardless of whether significant enhancement of SM ability is observed. The sample consisted of young adults (21 to 24 years of age) without regard for diagnostic status.

Consistent with previous studies (Gibson et al., 2011; Gibson, Kronenberger et al., 2012), serial position effects and estimates of PM and SM abilities were derived from performance on verbal and spatial IFR tasks. Of greatest importance, both the simple span and complex span training conditions produced significant (and equal) improvements in the SM component of WM capacity ($d = 0.868$ and $d = 0.691$, respectively). Likewise, as expected, both the simple span and complex span training conditions produced significant (and equal) improvements in the PM component of WM capacity ($d = .813$ and $d = .880$, respectively). Thus, the lower recall accuracy threshold had the intended effect on the SM component of WM capacity, and did not interact with exercise type. The present findings are thus consistent with the notion that span tasks may target the SM component when list length exceeds the capacity of PM (Unsworth and Engle, 2007b).

Stronger evidence for this conclusion was also sought by comparing the two training conditions used in this study with the two corresponding training conditions used in Gibson, Kronenberger et al.'s (2012) study. These latter two training conditions can be considered appropriate controls for the SM component of WM capacity because the active component of these two training conditions failed to have any effect on the SM component. A two-way ANCOVA was conducted with post-training estimates of SM ability serving as the dependent variable, and with training condition (simple span vs. complex span) and experiment (100% recall accuracy threshold vs. lower recall accuracy threshold) serving as the two between-subjects factors. Pre-training estimates of SM ability served as the covariate. As expected, there was a significant main effect of experiment, $F(1,75) = 13.19$, $p < .002$, $\eta_p^2 = .15$, indicating that individuals recalled significantly more items from SM after training with a lower recall accuracy threshold (adjusted $M = 2.83$ items) than after training with a 100% recall accuracy threshold (adjusted $M = 2.24$ items)—a 26% increase in performance—even after controlling for pre-existing differences in SM ability between the two experiments. In contrast, neither the main effect of training condition, nor the training condition X experiment interaction approached significance, $F < 1$. These findings are important because they suggest that the improvement in SM ability observed following training can be attributed to the use of a lower recall accuracy threshold. Although decreasing the recall accuracy threshold is subtle in absolute physical terms, this decrease can have important functional consequences because it elicits more retrieval from SM before recall is terminated on any given training trial, and it also allows the length of the training lists to be determined more by the limitations of the SM component than by the limitations of the PM component.

One of the potential virtues of using Gibson, Kronenberger et al.'s (2012) two training conditions as the control conditions in the present study is that these are active treatment conditions. As such, the active component of these training conditions may induce the same expectations for improvement as the lower recall accuracy threshold conditions while also having no effect on the SM component. However, lowering the recall accuracy threshold may increase motivation because individuals will tend to experience more success when they train using a lower recall accuracy threshold than when they train using a 100% recall accuracy threshold. Therefore, it is worth considering whether the significant enhancement of SM ability observed in the present study relative to Gibson, Kronenberger et al.'s (2012) study might be due to differences in motivation rather than differences in recall accuracy threshold.

If an alternative explanation based on motivation is viable, then it is reasonable to expect that the positive effects of motivation should extend to both the PM and SM components of WM capacity. Accordingly, a two-way ANCOVA was also conducted with post-training estimates of PM capacity serving as the dependent variable, and with training condition (simple span and complex span) and experiment (100% recall accuracy threshold vs. lower recall accuracy threshold) serving as the two between-subjects factors. Pre-training estimates of PM capacity served as the covariate. However, none of the effects approached significance (all p's > .10). This finding suggests that individuals recalled the same number of items from PM regardless of whether they were exposed to a lower recall accuracy threshold (adjusted $M = 3.08$ items) or to a 100% recall accuracy threshold (adjusted $M = 2.91$ items) during training. Altogether, the findings suggest that only the SM component of WM capacity benefitted from the lower recall accuracy threshold which is inconsistent with the alternative motivational account.

CONCLUSION

A large body of research has established that individual differences in WM capacity can constrain reasoning and learning abilities in both children and adults. Unfortunately, recent attempts to improve these outcomes by intensive training of WM capacity have proven to be unsuccessful, perhaps because these training regimens have not targeted the proper theoretical mechanisms. According to the dual-component theory of WM capacity, there are two critical components of WM capacity—the PM component and the SM component—that account for the relation between WM capacity and higher-level cognitive abilities. Within this context, we have hypothesized that improvements in student outcomes may be moderated by the number of WM components that are targeted and enhanced by a training regimen. In other words, the most potent training regimen would be one that could target and enhance both components of WM capacity. Consistent with this hypothesis, we have shown that existing span-based WM training regiments only appear to target the PM component of WM capacity (Gibson et al., 2011), which may explain why these regimens have proven to be relatively ineffective. In order to create a potentially more potent training regimen, we have provided preliminary evidence that the SM component of WM capacity can also be targeted and enhanced by decreasing the recall accuracy threshold; in contrast, manipulations of exercise type have not had the intended effect on the SM component (Gibson, Kronenberger et al., 2012). With this newly-created training regimen in hand, we are now in a position to provide a stronger test of the moderation hypothesis. Exploration of this hypothesis will not only serve to identify and develop a potentially beneficial educational intervention, it will also contribute to theory building regardless of whether the outcome is positive or negative.

REFERENCES

Alloway, T. P., Elliott, J., and Place, M. (2010). Investigating the relationship between attention and working memory in clinical and community samples. *Child Neuropsychology,* 16(3), 242-254.

Alloway, T. P., Gathercole, S. E., and Elliott, J. (2010). Examining the link between working memory behaviour and academic attainment in children with ADHD. *Developmental Medicine and Child Neurology,* 52(7), 632-636.

Alloway, T. P., Gathercole, S. E., Kirkwood, H., and Elliott, J. (2009). The cognitive and behavioral characteristics of children with low working memory. *Child Development,* 80(2), 606-621.

Archibald, L., and Joanisse, M. F. (2009). On the sensitivity and specificity of nonword repetition and sentence recall to language and memory impairments in children. *Journal of Speech, Language, and Hearing Research,* 52(4), 899-914.

Buschkuehl, M., and Jaeggi, S.M. (2010). Improving intelligence: A literature review. *Swiss Medical Weekly,* 140, 266-272.

Bull, R., and Scherif, G. (2001). Executive functioning as a predictor of children's mathematics ability: Inhibition, switching, and working memory. *Developmental Neuropsychology,* 19, 273-293.

Cain, K., and Oakhill, J. (2006). Profiles of children with specific reading comprehension difficulties. *British Journal of Educational Psychology,* 76, 683-696.

Chein, J.M., and Morrison, A.B. (2010). Expanding the mind's workspace: Training and transfer effects with a complex working memory span task. *Psychonomic Bulletin and Review,* 17, 193-199.

Cowan, N. (1995). Attention and memory: An integrated framework. Oxford, England: Oxford University Press.

Cowan, N. (2001). The magic number 4 in short-term memory: A reconsideration of mental storage capacity. *Behavioral and Brain Sciences,* 24, 97-185.

Cowan, N., Elliott, E. M., Saults, J. S., Morey, C. C., Mattox, S., Hismjatullina, A., and Conway, A. R. A (2005). On the capacity of attention: Its estimation and its role in working memory and cognitive aptitudes. *Cognitive Psychology,* 51, 42-100.

Craik, F. I. M., and Birtwistle, J. (1971). Proactive inhibition in free recall. *Journal of Experimental Psychology,* 91, 120-123.

Dahlin, K. I. E. (2010). Effects of working memory training on reading in children with special needs. *Reading and Writing,* 1-13.

Daneman, M., and Carpenter, P. A. (1980). Individual differences in working memory and reading. *Journal of Verbal Learning and Verbal Behavior,* 19, 450–466.

Diamond, A., and Lee, K. (2011). Interventions shown to aid executive function development in children 4 to 12 years old. *Science,* 333, 959-964.

DiPerna, J. C., and Elliott, S. N. (2000). Academic competence evaluation scales. San Antonio, TX: Psychological Corporation.

DuPaul, G. J., Volpe, R. J., Jitendra, A. K., Lutz, J. G., Lorah, K. S., and Gruber, R. (2004). Elementary school students with AD/HD: Predictors of academic achievement. *Journal of School Psychology,* 42(4), 285-301.

Engel de Abreu, P. M. J., Conway, A. R. A., and Gathercole, S. E. (2010). Working memory and fluid intelligence in young children. *Intelligence,* 38, 552–561.

Engel de Abreu, P. M. J., Gathercole, S. E., and Martin, R. (2011). Disentangling the relationship between working memory and language: The roles of short-term storage and cognitive control. *Learning and Individual Differences,* 21, 569–574.

Engle, R. W., Carullo, J. J., and Collins, K. W. (1991). Individual differences in working memory for comprehension and following directions. *Journal of Educational Research,* 84(5), 253-262.

Engle, R. W., Tuholski, S. W., Laughlin, J. E., and Conway, A. R. A. (1999). Working memory, short-term memory and general fluid intelligence: A latent-variable approach. *Journal of Experimental Psychology: General,* 128, 309–331.

Gathercole, S. E., Brown, L., and Pickering, S. J. (2003). Working memory assessments at school entry as longitudinal predictors of national curriculum attainment levels. *Educational and Child Psychology,* 20(3), 109–122.

Gathercole, S. E., and Pickering, S. J. (2000). Assessment of working memory in 6- and 7-year-old children. *Journal of Educational Psychology,* 92, 377–390.

Geary, D. C., Hoard, M. K., Nugent, L., and Byrd-Craven, J. (2007). Strategy use, long-term memory, and working memory capacity. In D. B. Berch and M. M. M. Mazzocco (Eds.), Why is math so hard for some children: The nature and origins of mathematical learning difficulties and disabilities (pp. 83-105). Baltimore, MD: Paul H. Brookes.

Gibson, B. S., Gondoli, D. M., Flies, A. C., Dobrzenski, B. A., and Unsworth, N. (2010). Application of the dual-component model of working memory to ADHD. *Child Neuropsychology*, 16 60-79.

Gibson, B.S., Gondoli, D.M., Johnson, A.C., Steeger, C.M., and Morrissey, R.M. (2012). The future promise of Cogmed working memory training. *Journal of Applied Research in Memory and Cognition*, 1, 214-216.

Gibson, B.S., Gondoli, D.M., Johnson, A.C., Steeger, C.M., Dobrzenski, B.A., and Morrissey, R.A. (2011). Component Analysis of Verbal versus Spatial Working Memory Training in Adolescents with ADHD: A Randomized, Controlled Trial. *Child Neuropsychology*, 17, 546-563.

Gibson, B.S., Gondoli, D.M., Kronenberger, W.G., Johnson, A.C., Steeger,C.M., and Morrissey, R.A. (2013). Exploration of an adaptive training regimen that can target the secondary memory component of working memory capacity. *Memory & Cognition*, DOI 10.3758/s13421-013-0295-8Gibson, B.S., Kronenberger, W.G., Gondoli, D.M., Johnson, A.C., Morrissey, R.M., and Steeger, C.M. (2012). Component analysis of simple span vs. complex span adaptive working memory exercises: A randomized, controlled trial. *Journal of Applied Research in Memory and Cognition*, 1, 179-184.

Holmes, J., Gathercole, S. E., and Dunning, D. L. (2009). Adaptive training leads to sustained enhancement of poor working memory in children. *Developmental Science*, 12(4), F9-F15.

Holmes, J., Gathercole, S.E., Place, M., Dunning, D.L., Hilton, K.A., and Elliott, J.G. (2010). Working memory deficits can be overcome: Impacts of training and medication on working memory in children with ADHD. *Applied Cognitive Psychology*, 24, 827-836.

Jaeggi, S. M, Buschkuehl, M., Jonides, J., and Perrig, W. J. (2008). Improved fluid intelligence with training on working memory. *Proceedings of the National Academy of Sciences of the United States of America*, 105, 6829-6833.

Jaeggi, S. M, Buschkuehl, M., Jonides, J., and Shah, P. (2011). Short- and long-term benefits of cognitive training. *Proceedings of the National Academy of Sciences of the United States of America*, 108, 10081-10086.

Kane, M. J., and Engle, R. W. (2002). The role of prefrontal cortex in working-memory capacity, executive attention, and general fluid intelligence: An individual-differences perspective. *Psychonomic Bulletin and Review*, 9, 637-671.

Kane, M. J., Hambrick, D. Z., Tulholski, S. W., Wilhelm, O., Payne, T. W., and Engle, R. W. (2004). The generality of working memory capacity: A latent-variable approach to verbal and visuospatial memory span and reasoning. *Journal of Experimental Psychology: General*, 133, 189-217.

Klingberg, T. (2010). Training and plasticity of working memory. *Trends in Cognition Sciences*, 14, 317-324.

Klingberg, T., Fernell, E., Olesen, P. J., Johnson, M., Gustafsson, P., Dahlström, K., and Westerberg, H. (2005). Computerized training of working memory in children with ADHD-a randomized, controlled trial. *Journal of the American Academy of Child and Adolescent Psychiatry*, 44(2), 177-186.

Kyttälä, M., Aunio, P., and Hautamäki, J. (2010). Working memory resources in young children with mathematical difficulties. *Scandinavian Journal of Psychology*, 51(1), 1-15.

Loo, S. K., Humphrey, L. A., Tapio, T., Moilanen, I. K., McGough, J. J., McCracken, J. T., and Smalley, S. L. (2007). Executive functioning among Finnish adolescents with attention-deficit/hyperactivity disorder. *Journal of the American Academy of Child and Adolescent Psychiatry,* 46(12), 1594-1604.

Melby-Lervåg, M., and Hulme, C. (2012). Is Working Memory Training Effective? A Meta-Analytic Review. Developmental Psychology. Advance online publication. doi: 10.1037/a0028228

Mezzacappa, E., and Buckner, J. C. (2010) Working memory training for children with attention problems or hyperactivity: A school-based pilot study. *School Mental Health*, 2, 202-208.

Miyake, A., Friedman, N. P., Rettinger, D. A., Shah, P., and Hegarty, M. (2001). How are visuospatial working memory, executive functioning, and spatial abilities related? A latent-variable analysis. *Journal of Experimental Psychology: General,* 130, 621-640.

Mogle, J. A., Lovett, B. J., Stawski, R. S., and Sliwinski, M. J. (2008). What's so special about working memory? An examination of the relationships among working memory, secondary memory, and fluid intelligence. *Psychological Science*, 19, 1071-1077.

Montgomery, J. W., Magimairaj, B. M., and Finney, M. C. (2010). Working memory and specific language impairment: An update on the relation and perspectives on assessment and treatment. *American Journal of Speech-Language Pathology,* 19(1), 78-94.

Morrison, A.B., and Chein, J.M. (2011). Does working memory training work? The promise and challenges of enhancing cognition by training working memory. *Psychonomic Bulletin and Review*, 18, 46-60.

Oberauer, (2005). The measurement of working memory capacity. In O. Wilhelm and R.W. Engle (Eds.), Understanding and measuring intelligence (pp. 393- 408). Thousand Oaks, CA: Sage.

Redick, T.S., Shipstead, Z., Harrison, T.L., Hicks, K.L., Fried, D.E., Hambrick, D.Z., Kane, M.J., and Engle, R.W. (2012). No evidence of intelligence improvement after working memory training: A randomized, placebo-controlled study. *Journal of Experimental Psychology: General*, 133, 189-217.

Shah, P., and Miyake, A. (1996). The separability of working memory resources for spatial thinking and language processing: An individual differences approach. *Journal of Experimental Psychology General,* 125, 4-27.

Shipstead, Z., Hicks, K.L., and Engle, R.W. (2012). Cogmed Working Memory Training: Does the Evidence Support the Claims? *Journal of Applied Research in Memory and Cognition.*

Shipstead, Z., Redick, T.S., and Engle, R.W. (2010). Does working memory training generalize? *Psychologica Belgica*, 50, 245-276.

Shipstead, Z., Redick, T.S., and Engle, R.W. (2012). Is working memory training effective? Psychological Bulletin. Advance online publication. doi: 10.1037/a0027473.

Swanson, H. L., and Beebe-Frankenberger, M. (2004). The relationship between working memory and mathematical problem solving in children at risk and not at risk for math disabilities. *Journal of Education Psychology,* 96, 471–491.

Swanson, H. L., Zheng, X., and Jerman, O. (2009). Working memory, short-term memory, and reading disabilities: A selective meta-analysis of the literature. Journal of Learning Disabilities, 42, 260-287.

St Clair-Thompson, H., Stevens, R., Hunt, A., and Bolder, E. (2010). Improving children's working memory and classroom performance. *Educational Psychology,* 30(2), 203-219.

Thorell, L. B., Lindqvist, S., Nutley, S. B., Bohlin, G., and Klingberg, T. (2009). Training and transfer effects of executive functions in preschool children. *Developmental Science,* 12(1), 106-113.

Tulving, E., and Colotla, V.A. (1970). Free recall of trilingual lists. *Cognitive Psychology,* 1, 86-98.

Turner, M. L., and Engle, R. W. (1989). Is working memory capacity task dependent? *Journal of Memory and Language*, 28, 127-154.

Unsworth, N. (2007). Individual differences in working memory capacity and episodic retrieval: Examining the dynamics of delayed and continuous distractor free recall. *Journal of Experimental Psychology: Learning, Memory, and Cognition*, 33, 1020-1034.

Unsworth, N. (2009). Variation in working memory capacity, fluid intelligence, and episodic recall: A latent variable examination of differences in the dynamics of free recall. *Memory and Cognition*, 37, 837-849.

Unsworth, N., Brewer, G.A., and Spillers, G.J. (2009). There's more to the working memory capacity-fluid intelligence relationship than just secondary memory. *Psychonomic Bulletin and Review*, 16, 931-937.

Unsworth, N., and Engle, R. W. (2006). Simple and complex memory spans and their relation to fluid abilities: Evidence from list-length effects. *Journal of Memory and Language*, 54, 68-80.

Unsworth, N., and Engle, R.W. (2007a). The nature of individual differences in working memory capacity: Active maintenance in primary memory and controlled search from secondary memory. *Psychological Review*, 114, 104-132.

Unsworth, N., and Engle, R. W. (2007b). On the division of short-term and working memory: An examination of simple and complex span and their relation to higher order abilities. *Psychological Bulletin*, 133, 1038-1066.

Unsworth, N., and Spillers, G.J. (2010). Working memory capacity: Attentional control, secondary memory, or both? A direct test of the dual-component model. *Journal of Memory and Language*, 62, 392-406.

Willcutt, E. G., Doyle, A. E., Nigg, J. T., Faraone, S. V., and Pennington, B. F. (2005). Validity of the executive function theory of attention-deficit/hyperactivity disorder: A meta-analytic review. *Biological Psychiatry,* 57(11), 1336-1346.

INDEX

A

academic learning, 29
academic performance, 3, 27, 176, 179, 188
academic problems, 202
access, 58, 64, 117, 159
acid, 149
active type, 109
adaptations, 187
adaptive functioning, 203
ADHD, 3, 12, 15, 43, 44, 45, 51, 175, 176, 177, 178, 179, 180, 181, 182, 183, 184, 185, 186, 187, 188, 189, 190, 191, 192, 193, 194, 195, 196, 197, 198, 199, 203, 208, 213, 215
ADHD symptoms, 185, 188, 189, 190, 203
adolescents, 53, 142, 176, 177, 180, 181, 184, 187, 189, 191, 193, 194, 197, 198, 208, 209, 215
adult education, 195
adulthood, 1, 10, 41, 50, 70, 86, 175, 191
adults, 13, 16, 33, 40, 41, 43, 50, 51, 53, 61, 64, 65, 68, 69, 70, 75, 76, 78, 83, 87, 88, 89, 90, 91, 92, 93, 94, 95, 138, 146, 151, 152, 153, 154, 177, 178, 181, 188, 189, 191, 193, 196, 197, 198, 199, 202, 203, 213
African Americans, 112
age, 1, 4, 5, 7, 8, 9, 10, 11, 14, 15, 17, 18, 19, 20, 23, 26, 29, 31, 32, 33, 35, 36, 37, 38, 39, 40, 41, 42, 43, 44, 45, 47, 48, 49, 50, 52, 53, 61, 62, 63, 67, 69, 70, 72, 73, 83, 86, 87, 88, 89, 90, 91, 92, 93, 94, 95, 96, 103, 128, 135, 138, 139, 142, 143, 146, 150, 153, 166, 177, 196, 198, 208, 209, 211
aging, 52, 83, 88, 93, 94, 95, 96, 128, 143
aging process, 91, 92
agonist, 149
alertness, 97, 99, 100, 101, 102, 103, 104, 105, 107, 108, 109, 110, 111
algorithm, 210
alpha activity, 174

alternative hypothesis, 50
alters, 154
American Psychiatric Association, 175
American Psychological Association, 154
amnesia, 91, 152
amygdala, 126, 149, 150, 154
anatomy, 41
anger, 197
animal learning, 145
ANOVA, 35, 36, 37, 38, 166, 167
anterior cingulate cortex, 116
anxiety, 103, 115, 118, 119, 123, 124, 127, 128, 129, 135, 143, 178, 186, 192
aptitude, 142, 202, 207
arithmetic, 3, 4, 10, 14, 97, 103, 105, 111, 203
arousal, 41, 94, 103, 113, 117, 122, 125, 128
articulation, 56, 64
aspartate, 178
assessment, 6, 11, 15, 20, 78, 84, 173, 185, 186, 188, 216
asymmetry, 174
Attention Deficit Hyperactivity Disorder, x, 43, 44, 175, 194, 197
attention to task, 58
attentional blink, 125, 127, 129
attentional disengagement, 69
attribution, 98
audition, 83, 123, 124, 125
autism, 51, 52, 53, 146, 193, 202
automatic processes, 142
automaticity, 85, 139, 141
automatisms, 98
autonomic nervous system, 112
awareness, 34, 46, 47, 98, 126, 132, 136

B

basal ganglia, 44, 189

base, 29, 30, 190
basic research, 146, 152
behavior of children, 24, 26, 30
behavioral assessment, 48
behaviors, 19, 21, 23, 26, 28, 154, 176, 185, 197, 202
beneficial effect, 190
benefits, 1, 146, 147, 150, 151, 160, 170, 184, 186, 189, 215
bias, 115, 118, 119, 120, 123, 124, 125, 127
bilateral, 157
binding, 75, 81, 83, 84, 90, 93, 94, 96, 174
biochemistry, 41
bipolar disorder, 80
blood, 103, 158
blood flow, 158
blood pressure, 103
borderline personality disorder, 193
bottom-up, 39
brain, 44, 50, 84, 92, 103, 126, 128, 135, 149, 150, 160, 161, 172, 173, 174, 189, 191, 193, 196, 199
brain activity, 103, 173
brain functions, 160
brothers, 197

C

cardiovascular disease, 112
case study, 91
catalyst, 148
central executive, 1, 2, 3, 4, 5, 6, 10, 18, 44, 58, 67, 84, 86, 116, 119, 125, 135, 159, 176, 177, 192, 197
central nervous system, 150, 172
cerebellum, 189
cerebral blood flow, 158
challenges, 165, 168, 216
Child Behavior Checklist, 183
child development, 44, 56, 58, 60, 61, 64, 65, 68, 71, 73, 74
childhood, 1, 3, 4, 9, 10, 11, 12, 41, 43, 44, 50, 51, 52, 66, 67, 68, 69, 70, 74, 75, 76, 79, 175, 195, 198
children, 1, 2, 3, 4, 5, 6, 7, 10, 11, 12, 13, 14, 15, 16, 17, 18, 19, 20, 21, 22, 23, 24, 25, 26, 27, 28, 29, 30, 31, 32, 33, 34, 35, 36, 37, 38, 39, 40, 41, 43, 44, 45, 47, 50, 51, 53, 55, 56, 60, 61, 63, 64, 65, 66, 67, 68, 69, 70, 71, 72, 73, 76, 77, 78, 79, 80, 138, 142, 143, 145, 146, 147, 151, 152, 153, 155, 175, 177, 179, 180, 181, 184, 185, 186, 187, 189, 191, 192, 193, 194, 195, 196, 197, 198, 202, 203, 213, 214, 215, 216
chunking, 64, 68, 70
clarity, 36

classes, 17, 19, 20, 21, 22, 23, 25, 26, 27, 28
classical conditioning, 149
classification, 177
classroom, 2, 6, 12, 18, 19, 21, 23, 26, 27, 29, 34, 35, 178, 185, 193, 197, 202, 216
clinical disorders, 176
close relationships, 4, 10, 11, 12
coding, 21, 67, 75, 77, 116
Cogmed, 180, 181, 182, 183, 184, 185, 187, 190, 208, 209, 215, 216
cognition, 5, 15, 44, 52, 76, 77, 78, 98, 115, 116, 117, 125, 154, 174, 175, 176, 181, 184, 190, 198, 206, 216
cognitive abilities, 73, 116, 188, 201, 206, 213
cognitive ability, 53
cognitive deficits, 172, 179
cognitive development, 51, 84
cognitive effort, 99, 100, 104, 105, 110
cognitive flexibility, 184
cognitive function, 12, 44, 50, 179, 180, 184, 188
cognitive load, 65, 66, 97, 98, 99, 100, 102, 103, 106, 107, 108, 110, 111, 112
cognitive map, 132, 133, 136
cognitive process, 32, 40, 42, 44, 83, 86, 92, 98, 105, 117, 176, 186
cognitive processing, 105, 117
cognitive psychology, 110, 135
cognitive remediation, 175, 179, 190
cognitive science, 52
cognitive skills, 190
cognitive system, 44, 98
cognitive tasks, 55, 166, 179
coherence, 158, 171
college students, 194, 197
color, 45, 77, 83, 84, 85, 86, 87, 89, 90, 91, 95, 120, 121, 139
combined effect, 198
commercial, 108
communication, 151, 153, 202
community, 45, 177, 213
comorbidity, 186, 194
competition, 15, 41
complexity, 61, 66, 68, 70, 74, 80
comprehension, 3, 5, 20, 41, 116, 127, 214
compression, 70, 74
computer, 6, 20, 46, 100, 118, 161, 162, 164, 173, 180, 187, 193
conception, 55, 203
conceptual model, 53, 197
conditioning, 149, 154
configuration, 61, 68
conflict, 95, 131, 139
connectivity, 157, 160, 170, 171, 173, 189, 196, 199

conscious awareness, 44
consciousness, 125
consensus, 116, 201
consent, 5
consolidation, 80, 120, 121, 124, 127, 129, 157, 158, 164, 171
constant rate, 125
continuum model, 62
contradiction, 44
control condition, 147, 189, 209, 210, 212
control group, 179, 181, 185, 187, 203
controlled trials, 179
controversial, 116, 138, 190
controversies, 75
correlation, 5, 7, 8, 33, 43, 48, 50, 51, 52, 99, 167, 170, 206
correlation analysis, 51
correlations, 1, 7, 8, 9, 18, 31, 51
Corsi Blocks task, 63
cortex, 72, 89, 158, 171
cortical neurons, 199
cost, 85, 118
cross-national, 193
cues, 68, 70, 120, 121, 126, 128, 149, 162, 164, 205
cultural practices, 139
curriculum, 3, 4, 7, 10, 11, 14, 18, 19, 29, 202, 214
cycles, 164

D

data collection, 51, 172
data set, 50
decay, 64, 66, 73, 86, 120
decoding, 2, 203
deficiencies, 191
deficiency, 154
deficit, 53, 87, 88, 89, 90, 91, 94, 95, 175, 176, 177, 180, 186, 191, 192, 193, 194, 195, 196, 197, 198, 199, 202, 216, 217
degenerate, 90
degradation, 91
dementia, 91, 150, 153
dependent variable, 34, 35, 36, 38, 67, 165, 212
depolarization, 160
depression, 95
deprivation, 152
depth, 35, 40, 158
detection, 43, 45, 71, 72, 120, 121
developed countries, 152
developing children, viii, xi, 45, 179, 180
developmental change, 11, 63, 72, 74
developmental disorder, 19, 175
developmental dyslexia, 15

developmental milestones, 138
developmental process, 137
Diagnostic and Statistical Manual of Mental Disorders, 191
directives, 23
disability, 14, 202
discomfort, 165
discrimination, 100, 120, 121, 145, 146, 148, 149, 152, 153, 154
discrimination learning, 146, 148, 152
disorder, 51, 53, 161, 175, 176, 177, 178, 179, 184, 186, 188, 189, 191, 192, 193, 194, 195, 196, 197, 198, 199, 202, 216, 217
displacement, 207
dissociation, 52, 149
distracters, 6
distribution, 168, 169, 176
diversity, 42, 64
dogs, 146
DOI, 75, 76, 215
dominance, 4, 67, 115, 122, 123, 124, 125, 126
DOP, 146, 147, 148, 150, 151, 152
dopamine, 190, 191, 192, 196
dopaminergic, 189
dorsolateral prefrontal cortex (DLPFC), x, 116, 157, 158, 172, 194
Down syndrome, 152
drawing, 38, 66, 142
drug treatment, 197
drugs, 154
dual task, 4, 64, 68
dyslexia, 3, 15, 198

E

education, 29, 90, 166
educational attainment, 5, 14, 29
educational psychology, 29, 97
educational research, 111
educational system, 209
educators, 12
EEG, 158, 160, 163, 170, 171, 172, 173, 174
egg, 31, 32, 34, 35, 38, 39, 40
electrodes, 158, 163
electroencephalography, 160
emission, 158
emotion, 111, 115, 116, 117, 126, 128, 154
emotional stimuli, 129
empirical studies, 180, 203, 205
employment, 55
encoding, 61, 62, 69, 71, 74, 76, 79, 80, 83, 85, 86, 88, 90, 94, 134, 157, 158, 160, 168, 170, 173, 174, 177

encouragement, 187
entorhinal cortex, 90, 158
environment, 85, 131, 132, 133, 134, 135, 137, 138, 140, 143, 178, 187, 192
environments, 131, 136, 138, 139, 142, 143, 178
epilepsy, 161
episodic memory, 86, 87, 141, 157, 160, 165, 167, 168, 170, 171, 174, 190
equipment, 107
ergonomics, 97, 110
ethnicity, 5
event-related potential, 140
everyday life, 115, 116, 122, 131, 178, 181, 184, 188
evidence, 11, 19, 27, 32, 38, 44, 47, 50, 52, 56, 58, 63, 64, 65, 67, 68, 69, 79, 85, 88, 89, 91, 94, 95, 118, 121, 123, 129, 131, 132, 136, 145, 172, 173, 177, 178, 179, 184, 186, 189, 190, 192, 201, 203, 205, 209, 210, 211, 213, 216
evil, 180
evolution, 176
excitability, 172, 174
exclusion, 161, 186
execution, 35, 36, 38, 39, 40, 98, 102
executive function(s), 4, 13, 32, 33, 41, 42, 94, 98, 179, 184, 185, 186, 187, 189, 191, 192, 193, 195, 196, 199, 202, 214, 216, 217
executive functioning, 32, 94, 186, 191, 192, 196, 216
executive processes, 67, 128
exercise, 211, 213
experimental condition, 103, 104
expertise, 68, 80, 111
explicit memory, 158, 170, 171, 173
exposure, 4, 133, 134, 139, 170

F

factor analysis, 207
family conflict, 108, 109
fidelity, 71, 72, 73
fixation, 45, 162, 163
flight, 107
fluid, 14, 73, 77, 127, 173, 202, 203, 205, 206, 207, 214, 215, 216, 217
fluid intelligence, 14, 73, 77, 127, 173, 203, 206, 207, 214, 215, 216, 217
fMRI, 53, 80, 94, 142, 150, 153, 173, 189
food, 111, 112, 147, 148
forebrain, 158
foreign language, 3
formation, 88, 91, 149
formula, 46
fragility, 86, 90, 96

fragments, 46
framing, 64
free recall, 111, 160, 165, 169, 170, 171, 207, 214, 217
freedom, 139
frontal lobe, 41, 42, 128
functional architecture, 71, 74
functional imaging, 126

G

GABA, 149
gender differences, 161
gender gap, 139
general intelligence, 4
general school achievement, vii
generalizability, 190
genes, 190
genotype, 192
germane load, ix, 97, 98, 99, 102, 110
goal-directed behavior, 176
goal-oriented behaviour, 32
goal-setting, 32
grades, 179
graph, 61
grouping, 68
growth, 13
guessing, 164, 166
guidance, 134, 136, 137, 140, 143

H

handedness, 161, 166, 173
head injury, 186
health, 34, 107, 112
heart rate, 99, 100, 102, 103, 104, 107, 111
hemisphere, 123
heterogeneity, 180, 188
hippocampus, 63, 89, 90, 91, 93, 95, 96, 143, 149, 155, 158, 165, 171
history, 161
homogeneity, 166
hub, 84, 176
human, 53, 96, 123, 124, 126, 129, 145, 146, 147, 148, 149, 150, 153, 154, 157, 158, 171, 172, 174, 176
human brain, 53
hyperactivity, 51, 53, 175, 178, 181, 184, 185, 191, 192, 193, 194, 195, 196, 197, 198, 199, 202, 216, 217
hypothesis, 33, 40, 43, 44, 50, 51, 87, 95, 99, 102, 103, 159, 168, 190, 201, 211, 213

I

ideal, 152
identification, 56, 202
identity, 86, 87, 112
idiosyncratic, 92
image(s), 56, 66, 72, 112, 138
imagery, 78, 117
impairments, 13, 93, 149, 150, 154, 172, 187, 195, 196, 213
implants, 161
implicit knowledge, 53
implicit learning, 53
implicit memory, 149, 154, 171
improvements, 50, 63, 73, 145, 178, 179, 180, 181, 185, 186, 188, 201, 203, 211, 213
impulsivity, 175, 178, 185, 197
inattention, 51, 175, 178, 184, 185
income, 5
independent variable, 35, 36, 37, 38
indirect measure, 52
individual character, 98, 103
individual characteristics, 98, 103
individual differences, 3, 13, 19, 22, 23, 27, 28, 52, 53, 60, 64, 78, 79, 80, 131, 133, 134, 135, 136, 137, 139, 140, 141, 142, 190, 197, 202, 205, 206, 213, 216, 217
individuals, 14, 118, 120, 131, 133, 134, 135, 136, 139, 140, 143, 153, 159, 178, 180, 186, 198, 204, 205, 206, 207, 212
induction, 168, 171
infancy, 44, 76
infants, 53
information processing, 123, 124, 125
informed consent, 45, 161
inhibition, viii, 31, 32, 33, 35, 38, 39, 40, 42, 66, 69, 116, 129, 154, 159, 176, 184, 185, 186, 191, 192, 194, 197, 214
inhibition, 13, 16, 31, 34, 213
initiation, 184
injuries, 176
instructional design, 111, 112, 113
integration, 68, 95, 135, 172
integrity, 160
intelligence, 15, 73, 76, 116, 155, 213, 216
interdependence, 170
interface, 2, 56, 92
interference, 2, 15, 64, 66, 68, 71, 85, 92, 93, 96, 115, 117, 118, 119, 120, 121, 122, 123, 124, 125, 126, 127, 135, 137, 143, 159, 162, 184, 206
internal consistency, 197
interneurons, 194
intervention, 27, 29, 193, 208, 213

investment, 103, 105, 180
IQ scores, 207
isolation, 107
issues, 29, 56, 67, 69, 72, 97, 110, 132, 175
iteration, 181

J

JJapanese, 17, 20, 21, 22, 23, 24, 25, 26, 27, 28, 29, 30, 127, 128
job characteristics, 107, 108, 110, 111, 112
job performance, 108, 190
job strain, 107

K

knowledge acquisition, 143

L

labeling, 153
laboratory studies, 99, 107
language impairment, 3, 13, 194, 216
language processing, 216
latency, 108, 205
lead, 19, 23, 24, 69, 70, 107, 137, 145, 160, 204
learning, 1, 3, 4, 12, 13, 14, 15, 17, 18, 19, 21, 22, 23, 26, 28, 29, 30, 43, 44, 46, 47, 48, 49, 50, 51, 52, 53, 75, 78, 88, 93, 94, 98, 99, 111, 112, 113, 131, 133, 134, 136, 137, 139, 141, 142, 145, 146, 148, 149, 152, 153, 154, 155, 158, 160, 161, 164, 165, 168, 170, 172, 173, 177, 186, 196, 198, 201, 202, 203, 204, 206, 213, 214
learning behavior, 158
learning difficulties, 3, 15, 18, 214
learning disabilities, 161
learning process, 99, 155
learning task, 46, 137
lesions, 78, 149, 150, 154
light, 107, 108, 109, 145, 146, 148, 168
literacy, 3, 4, 11, 19
localization, 153, 164, 172
location information, 87
locus, 129
longevity, 190
longitudinal study, 52
long-term memory, 2, 3, 18, 68, 78, 83, 84, 91, 94, 115, 117, 176, 214

M

magnetic resonance, 189
magnetic resonance imaging, 189
magnitude, 44, 51
major depressive disorder, 128
majority, 58, 60, 83, 85, 86, 178, 187, 188
man, 78, 151
management, 116
manipulation, 31, 61, 66, 69, 70, 88, 89, 115, 116, 120, 124, 157, 159, 160, 161, 164, 166, 171, 208
manufacturing, 195
mapping, 64, 172
matching-to-sample, 146, 152, 153
materials, 88
mathematics, 3, 4, 5, 6, 7, 8, 10, 12, 15, 17, 18, 19, 21, 22, 23, 24, 25, 26, 27, 28, 29, 213
matrix, 6, 60, 61, 62, 70
measurement, 15, 16, 61, 95, 112, 173, 185, 188, 192, 193, 203, 206, 216
measurements, 163
media, 55, 190
medication, 84, 161, 178, 181, 187, 189, 194, 195, 198, 199, 215
memory capacity, 12, 15, 18, 19, 33, 41, 42, 46, 51, 53, 61, 69, 70, 90, 127, 135, 173, 196, 204, 215, 217
memory formation, 157, 172
memory function, 74
memory performance, 85, 86, 95, 100, 107, 145, 148, 150, 171, 173
memory processes, 13, 50, 56, 95, 135
memory retrieval, 158, 171
mental arithmetic, 2, 14, 97, 102, 103, 104, 105, 108, 109
mental health, 113
mental load, 103
mental processes, 103
mental retardation, 146
messages, 123
meta-analysis, 86, 87, 95, 96, 127, 139, 144, 179, 195, 197, 198, 203, 216
metabolic, 111
methodological principles, 97
methodology, 31, 33, 85, 90, 128
methylphenidate, 178, 179, 187, 189, 192, 196, 197, 198
models, 2, 4, 11, 75, 83, 84, 107, 108, 115, 116, 117, 150, 154, 158, 172, 176, 197
mood disorder, 186
motivation, 14, 29, 75, 93, 99, 111, 141, 180, 188, 193, 197, 210, 212
MTS, 146

multidimensional, 56, 113, 132, 176
multiple regression analyses, 100
multiple regression analysis, 100
multiplication, 15

N

naming, 16, 67, 70, 139
National Academy of Sciences, 191, 215
navigation, 131, 132, 133, 135, 137, 138, 141, 142, 143
navigation system, 137, 143
navigation tasks, ix
navigational abilities, 132, 138
negative outcomes, 175, 176
negative valence, 118, 119, 120, 121, 125
negativity, 120, 125, 127
neglect, 15, 41
neurodevelopmental disorders, 3, 44, 45, 51
neuroimaging, 190
neurons, 96, 160, 173
neurophysiology, 52
neuropsychology, 51, 126
neuroscience, 52, 53, 79, 172
neurosurgery, 78
neurotransmitter, 190
neutral, 115, 116, 117, 118, 119, 120, 121, 122, 123, 125, 126, 153
null, 170

O

oculomotor, 197
old age, 192
omission, 184
operations, 6
opportunities, 2, 19, 107, 139, 210
oscillation, 160, 173
oscillators, 160
oscillatory activity, 159
outpatient, 196
overlap, 2

P

pain, 165, 166, 173
panic disorder, 128
parallel, 74, 142
parental involvement, 187
parenting, 194
parents, 152, 184, 185, 187, 188, 190
parietal cortex, 129, 150

parietal lobe, 75, 189
participants, 2, 4, 5, 6, 11, 20, 32, 33, 35, 40, 41, 44, 45, 46, 47, 48, 68, 71, 72, 84, 85, 86, 90, 92, 97, 99, 100, 101, 102, 103, 104, 105, 108, 118, 121, 122, 123, 137, 140, 146, 149, 152, 157, 159, 160, 161, 162, 163, 164, 165, 166, 168, 169, 170, 185, 186, 187, 189, 190, 206
pathways, 160
pedagogical approaches, vii, 19, 23, 24, 26, 28
peer relationship, 48, 51
percentile, 18, 19, 202
permission, 20
personality, 103
personality traits, 103
phonological deficit, 15
phonological form, 83
photographs, 147
physiological, 154
pilot study, 185, 186, 187, 194, 196, 197, 216
pitch, 122
placebo, 187, 189, 216
plants, 161
plasticity, 160, 168, 190, 192, 195, 198, 215
polymorphisms, 190
population, 175
position effect, 208, 210, 211
positive correlation, 108, 135
positron, 174, 190
positron emission tomography, 174, 190
positron emission tomography (PET), 128, 158
prefrontal cortex, 14, 41, 89, 116, 126, 158, 160, 171, 173, 199, 215
pregnancy, 161
prematurity, 202
preparation, 28, 93, 147
preschool, 53, 217
preschool children, 217
preschoolers, 177, 195
presentation order, 162
prevalence rate, 175
primacy, 177
primary school, 27
primate, 194
priming, 70, 128
principles, 174
probability, 134, 147, 209, 210
probe, 69, 72, 84, 85, 162, 164
problem solving, 3, 10, 98, 112, 113, 154, 184, 186, 216
process control, 111
processing speed, 77, 194
proposition, 118
prototypes, 108

psychiatric disorders, 186
psychological processes, 99
psychological stress, 107
psychological stressors, 107
psychology, 14, 29, 53, 74, 75, 76, 77, 79, 80, 93, 99, 103, 111, 112, 141, 194
psychopathology, 53, 129, 197
psychosocial stress, 107, 109

Q

quality of life, 176, 191, 192
quantification, 185
questioning, 21
questionnaire, 108, 110, 161, 182

R

radius, 45
rating scale, 45, 184, 185, 188
reaction time, 6, 43, 44, 46, 47, 48, 49, 53, 151, 166, 183
reactions, 143
reading, 3, 5, 6, 7, 8, 9, 10, 12, 13, 15, 16, 18, 77, 98, 116, 127, 177, 183, 193, 194, 201, 202, 203, 206, 207, 214, 216
reading comprehension, 13, 15, 16, 77, 202, 203, 206, 214
reading disorder, 194
reasoning, 4, 116, 184, 186, 201, 202, 203, 204, 206, 213, 215
recall, 2, 4, 6, 7, 8, 9, 10, 11, 61, 63, 65, 67, 68, 76, 85, 86, 87, 89, 94, 97, 100, 111, 122, 123, 124, 125, 127, 157, 161, 163, 164, 165, 168, 169, 170, 171, 172, 204, 205, 206, 207, 210, 211, 212, 213, 217
recalling, 65, 164, 171, 207
recency effect, 68, 72
recognition, 46, 47, 61, 69, 74, 84, 85, 86, 87, 88, 89, 90, 93, 118, 119, 123, 126, 147, 148, 151, 152, 153, 154, 158, 161, 162, 164, 173
recognition test, 46
recombination, 87
recommendations, iv
reconstruction, 85
recovery, 204, 205
reference frame, 134, 137, 142
regression, 63, 100, 101
regression analysis, 100
rehabilitation, 84, 198, 199
reinforcement, 147, 152, 153, 154, 155
reinforcers, 147, 148

rejection, 88
relevance, 66, 120, 125, 160, 170, 171
reliability, 6, 15, 52, 61, 99, 197
remediation, 154, 171, 175, 179, 180, 181, 187, 188, 190, 196
replication, 67
requirements, 2, 53, 58, 110, 111
RES, 75
researchers, 4, 11, 63, 84, 116, 132, 149, 150, 201
resolution, 72, 80, 92, 103, 172
resources, 4, 10, 14, 15, 32, 40, 41, 55, 56, 58, 64, 67, 72, 73, 74, 80, 85, 94, 98, 100, 102, 103, 105, 108, 109, 110, 128, 132, 133, 135, 136, 137, 176, 215, 216
response, 6, 13, 15, 17, 25, 28, 31, 32, 33, 36, 39, 40, 41, 46, 48, 67, 87, 103, 108, 109, 111, 115, 119, 129, 138, 146, 147, 148, 159, 160, 162, 163, 164, 165, 184, 186, 190, 192, 194
response time, 36, 115, 119, 138, 184
responsiveness, 52
resting potential, 160
retention interval, 86, 158
retroactive interference, 68
rewards, 148, 150, 153
rhythm, 158, 160, 168, 171
right hemisphere, 63, 123
risk, 11, 202, 216
rotations, 137
routes, 132, 136, 137

S

safety, 98, 99, 110, 111
schizophrenia, 172, 173
scholastic attainment, vii, 1, 4, 8, 10, 11, 12
school, 2, 4, 5, 6, 14, 17, 19, 20, 26, 29, 34, 53, 166, 175, 179, 184, 185, 189, 190, 192, 196, 197, 208, 214, 216
school classroom, vii, 2, 197
schooling, 4, 29
science, 3, 4, 10, 18, 75, 76, 111, 112
scope, 58, 69, 70, 71, 74, 99, 180
scope of attention, 76
search terms, 180
seizure, 186
selective attention, 33, 41, 77, 121, 127
self-esteem, 18
self-regulation, 187, 191
semantic information, 115, 140
semantic processing, 139
semantics, 56, 64, 67, 68, 69, 70, 72, 74, 140
sensation, 164, 168
sensations, 168

sensitivity, 99, 159, 170, 213
sensory modalities, 176
sex, 133, 135, 139, 142, 144
sex differences, 139, 142, 144
sham, 160, 161, 164, 165, 166, 167, 168, 170, 171
shape, 57, 83, 84, 85, 86, 89, 90, 91, 112
short term memory, 55, 68, 74, 77, 184, 186
short-term memory, 2, 3, 4, 10, 14, 15, 16, 18, 20, 29, 52, 74, 77, 78, 79, 80, 85, 90, 93, 94, 95, 96, 98, 105, 116, 125, 127, 129, 146, 148, 152, 174, 182, 184, 186, 214, 216
showing, 26, 34, 40, 86, 91, 119, 134, 136, 138, 149
siblings, 177
side effects, 166
significance level, 167
simulation, 136
single test, 85
skin, 161, 164
sleep deprivation, 152, 153
social identity, 112
social problems, 195
social psychology, 112
social support, 108, 109
socioeconomic status, 177
software, 6, 161, 187
spatial ability, 138, 139, 143
spatial cognition, 131
spatial information, 2, 131, 133, 134, 135, 139, 140, 143, 174, 208
spatial learning, 172
spatial location, 63, 77, 83, 133, 139, 140
spatial memory, 63, 77, 79, 141, 143, 147
spatial orientation, 132
spatial processing, 56, 142
special education, 12
specialization, 50, 63, 193
species, 145, 146
speech, 126, 176
speed of processing, 11
spending, 36
sponge, 163
stability, 72, 185
standard deviation, 166
state, 99, 108, 148, 203
states, 160, 171
stereotypes, 112
stereotyping, 103
stimulant, 178, 189, 198, 199
stimulation, 157, 159, 161, 163, 164, 165, 166, 167, 168, 169, 170, 171, 172, 173, 174
stimulus, 45, 46, 65, 68, 71, 72, 83, 85, 91, 118, 119, 146, 147, 148, 149, 151, 153, 159, 162
stimulus information, 159

stochastic resonance, 160
storage, 2, 4, 5, 10, 11, 12, 14, 18, 40, 50, 56, 58, 64, 75, 76, 96, 98, 127, 159, 176, 177, 183, 207, 208, 214
strategy use, 88, 92, 137, 181
stress, 103, 107, 110, 111, 112
stress factors, 107
stressors, 107
structural equation modeling, 205
structure, 2, 6, 84, 92, 149
style, 48, 134, 140
subitizing, 118, 119
subitizing, 128
subtraction, 15
sucrose, 146, 147, 148
supervision, 98
suppression, 70, 129, 137
survey knowledge, 133
survival, 120, 125
symmetry, 56, 68, 79, 209
symptoms, 45, 175, 178, 180, 183, 184, 185, 188, 190, 203
synaptic plasticity, 194, 199
synchronization, 157, 158, 160, 161, 170, 171, 172, 173, 174
synchronize, 158, 171
syndrome, 146, 150, 153, 195
synthesis, 155

T

target, 6, 12, 21, 22, 23, 24, 26, 27, 67, 69, 70, 84, 87, 120, 121, 159, 161, 162, 164, 178, 201, 203, 204, 206, 209, 210, 211, 213, 215
target stimuli, 67, 69, 84
task conditions, 100, 105, 108
task demands, 40, 56, 58, 64, 73, 86, 98
task difficulty, 97, 98, 100, 101, 102, 103, 104, 105, 107, 108, 109, 110, 172, 177
task interference, 64, 143
task load, 207
task performance, 4, 33, 40, 42, 55, 56, 61, 64, 66, 67, 68, 69, 70, 71, 72, 73, 98, 99, 103, 108, 111, 135, 140, 207, 208
teachers, 14, 17, 18, 19, 21, 22, 23, 24, 25, 26, 27, 28, 29, 112, 184, 185, 188, 190
technician, 6
temperature, 99, 111
temporal lobe, 63, 78, 91, 94, 95
tension, 99, 100, 103, 107, 108, 109, 111
test scores, 7, 19
testing, 5, 6, 20, 87, 164, 183
textbook(s), 21, 24

texture, 71
thalamus, 189
theoretical approaches, ix, xi
therapy, 180
thesaurus, 173
thoughts, 32, 103
tics, 77
time periods, 117
time pressure, 97, 98, 100, 101, 102, 103, 104, 105, 107, 108, 109, 110
top-down, 170
toxicity, 150
trade, 47
trainees, 210
training, 12, 27, 46, 149, 151, 152, 153, 175, 180, 181, 184, 185, 186, 187, 188, 189, 190, 191, 192, 193, 194, 195, 196, 197, 198, 199, 201, 203, 204, 206, 207, 208, 209, 210, 211, 212, 213, 214, 215, 216
training block, 46
training programs, 186, 187, 190, 203, 209
trajectory, 61, 62
transactions, 144
transcranial alternating current stimulation (tACS), x, 157, 172
transmission, 113
traumatic brain injury, 202
treatment, 39, 53, 175, 178, 179, 180, 181, 187, 188, 189, 190, 192, 197, 198, 209, 212, 216
trial, 6, 45, 46, 108, 118, 121, 122, 146, 152, 154, 159, 162, 163, 180, 191, 193, 194, 195, 199, 206, 207, 210, 211, 212, 215
type II error, 50

U

updating, 40, 42, 94, 116, 118, 119, 121, 132, 143, 206, 208

V

valence, 115, 117, 118, 119, 120, 121, 122, 123, 124, 125, 127
variables, 88, 107, 144
variations, 96, 99, 100, 102, 104, 105, 111
varieties, 206
vein, 27, 83
verbal fluency, 64
vision, 75, 83, 124, 125, 161, 186
visual attention, 72, 127
visual environment, 83, 84
visual field, 52

visual images, 79, 176
visual stimuli, 120
visualization, 72, 73, 79
Visuo-spatial working memory, 142
vocabulary, 3, 10, 14, 15, 18, 29
vocalizations, 128
VPT, 61, 62, 63, 64, 68, 70, 71, 72

W

waking, 173
water, 147
wayfinding, 132
wear, 6
Wechsler Intelligence Scale, 183
word frequency, 117
word recognition, 118, 162
work environment, 107
workers, 98, 103, 107, 111, 193
working conditions, 107, 108
Working memory, 2, 3, 12, 13, 14, 15, 16, 18, 29, 41, 42, 46, 51, 53, 75, 77, 79, 80, 93, 94, 95, 98, 111, 126, 127, 128, 141, 172, 176, 189, 191, 192, 193, 194, 195, 196, 197, 198, 201, 214, 215, 216, 217
Working memory training, 189, 196, 216
workload, 98, 99, 110, 111, 134, 136, 137
workplace, 18, 176, 195
World Health Organization (WHO), 175, 193, 195
writing, 8, 9, 15, 214

Y

Y-axis, 168
yield, 187, 203
young adults, 32, 44, 50, 61, 83, 84, 85, 86, 91, 191, 198, 202, 211
young people, 151

RECEIVED

NOV 2 7 2013

GUELPH HUMBER LIBRARY
205 Humber College Blvd
Toronto, ON   M9W 5L7